Abandoned Railroads of Delmarva

DOUGLAS POORE

America Through Time®
Adding Color to American History

America Through Time is an imprint of Fonthill Media LLC
www.through-time.com
office@through-time.com

Published by Arcadia Publishing by arrangement with Fonthill Media LLC
For all general information, please contact Arcadia Publishing:
Telephone: 843-853-2070
Fax: 843-853-0044
E-mail: sales@arcadiapublishing.com
For customer service and orders:
Toll-Free 1-888-313-2665

www.arcadiapublishing.com

First published 2021

ISBN 978-1-63499-285-5

Typeset in Mrs Eaves XL Serif Narrow
Printed and bound in England

Contents

Acknowledgments

To my dad, thank you for my first model train. It turned a child's love of trains into a lifelong passion.

To all the train buffs that inspired and aided me: Jim Bowden, Keith Short, Robert Biddle, Ed Sharpe and so many more. Each of you gentleman has guided and encouraged me in this process.

To my wife Lisa: are you tired of trains yet?

To Kena and the rest of the Fonthill staff: thank you all for the guidance and encouragement.

To DFC A. Ortiz of Somerset County Sheriff's Office: thanks for just a warning for the stop sign I might have slowly rolled through! You rock!

To the folks at the Brown Dog in Cape Charles: great chocolate milkshake!

To all the museums, historical societies, Facebook group members and anyone else that provided photos and stories for this book: thank you. A special thank you to the Greater Harrington Historical Society. Any photo not marked is courtesy of this organization.

To all that read this book: I hope this look into history gives you an appreciation for the past. Do not rush by just going from here to there. Take the backroads, find a place new to stop and visit. Sit and listen, pick a flower, and give it to the one you love. Look at the butterflies.

Introduction

Abandoned, according to Webster's Dictionary, means deserted, forsaken, cast off, neglected, and empty. So many things in our country, local area, and even daily lives are abandoned. Some we outgrow, such as clothes, cars, furniture, and other material things. Most of these we cast aside without another thought given to their existence. But one shirt that was worn when an event happened seems to never leave us.

Sadly, people come and go in our lives, too. We do not often use the word abandoned for them, but rather say we grew apart. There is that rare occasion where someone just goes, leaving both happy and sad memories behind. The shirt left behind by the special person leaves a hole in our heart and soul. The memories can make us smile, but often with a tear. Usually, we look back at these things and times with fond memories. We reminisce about what once was. However, some of these elicit hurt and pain for us, reminding us of the loss of a loved one, a job, or an era in our past when life was simpler. Whether good or bad, these are the things that make us. We are a tapestry, made up of all these threads. Pull one and our entire story unravels. Such as it is with the history of anything that has ever existed, such as countries, buildings, cars, and so much more. We feel a connection with the things that were a daily part of our lives—the things that hold those special places in our hearts and memories.

For so many in America, the railroad is that memory. We reminisce about the romance of the railroad and watching as the trains passed us on their travels, loaded with people and goods to destinations all around the country. When the railroads began in America, most people had never traveled more than 5 miles from their home. Most goods and services came from nearby cities and towns. The distance would be a wagon ride of maybe a day. Some did travel by sail on the waterways. These journeys could be longer, but again, for most they only would hear the stories of those that lived this adventure. The railroad would change all of that. Travel that would take days now took hours. We began to travel for leisure, something never dreamed of in most people's lives. Goods were also now available to the masses. Clothing, food, and all kinds of items were now available in areas that had never tasted or used such products. When America first moved west, we only had what we could carry in a wagon. The train brought comfort items to the ever-expanding country. From its beginning, the railroad played a vital role in wartime as well. From moving goods and troops in the Civil War to conflicts of today, railroads have been at the forefront of our nation's war efforts.

The era just after the Civil War is considered the Golden Era of Railroads. Even the smallest towns and remote areas of the country would be serviced by rails. It took another conflict, World War II, to draw that era to a close. People began to travel more by car. The Interstate System came into being, allowing motorist to arrive at their destination by car quicker than by train. Goods would also begin to traverse the country on the ribbons of asphalt and concrete, reducing the need for cargo to travel by rail. By the 1960s, the railroad industry was struggling. Many had cut back on or eliminated passenger service. Freight trains that on some systems would be over 100 cars long, multiple times daily, would now see trains half that length and with less frequency. This all came to a head in the 1970s, with the bankruptcy of the mighty Pennsylvania Railroad. Once the king of the industry, the company was bleeding money at a staggering rate. The federal government had to step in to stop the complete shutdown of the industry.

Through various federal legislation, Conrail was created for freight service, and Amtrak for passenger service. The Staggers Rail Act of 1980 finally allowed railroads some breathing room. Able to be competitive with other industries that moved freight now, railroads began to make a comeback. Companies became profitable again, and lines were abandoned that were no longer economically feasible. Finally, in 1997, Conrail was returned to private ownership as the railroad industry had returned from near extinction.

As you ride the countryside, look closely and you can see remnants of what was. Old abandoned stations now serve a variety of roles, reborn as houses, offices, and museums. Rail beds today are roadways, bike paths, or sit abandoned with rails rusting away, and the occasional visitors stopping to take a photo and reminisce about what was once a part of their everyday life. This book pays homage to those memories, to a time when you could hear the clickety-clack across America, when women would not put laundry out to dry at certain times of the day because the soot from the smokestacks of the engine would only dirty it again. So, sit back in your favorite chair and enjoy this look back at a time when rails were king.

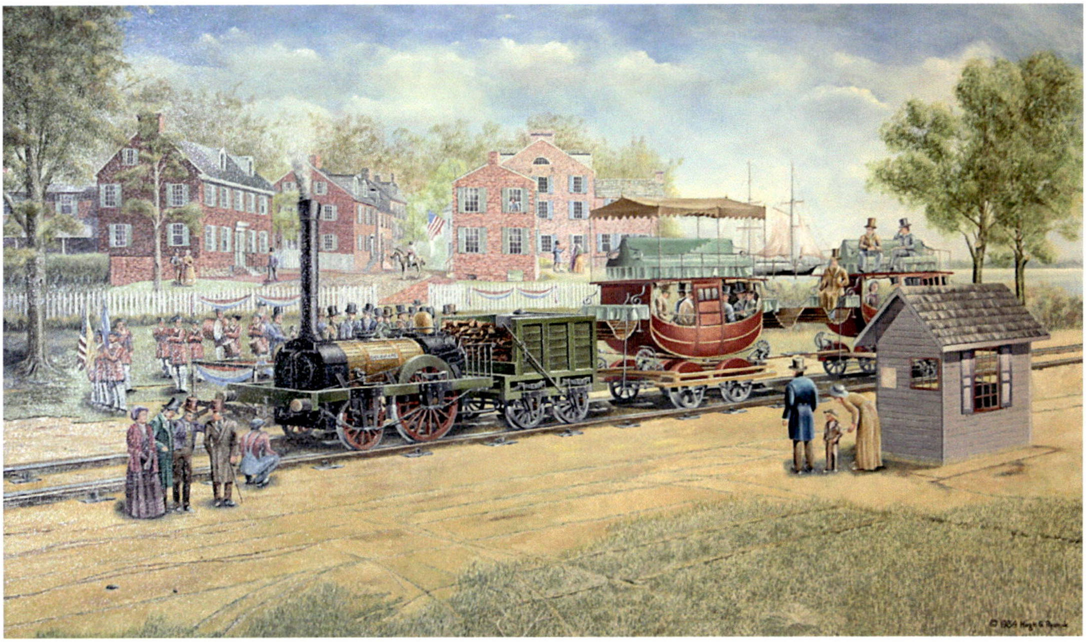

Painted in 1984 by Hugh Ryan Jr., the painting celebrates the opening of the New Castle and Frenchtown Railroad in 1832. The railroad began as a turnpike, and it was reorganized in 1830 to become one of America's first railroads, traveling from New Castle, Delaware, on the Delaware River, to Frenchtown Maryland, located on the Elk River.

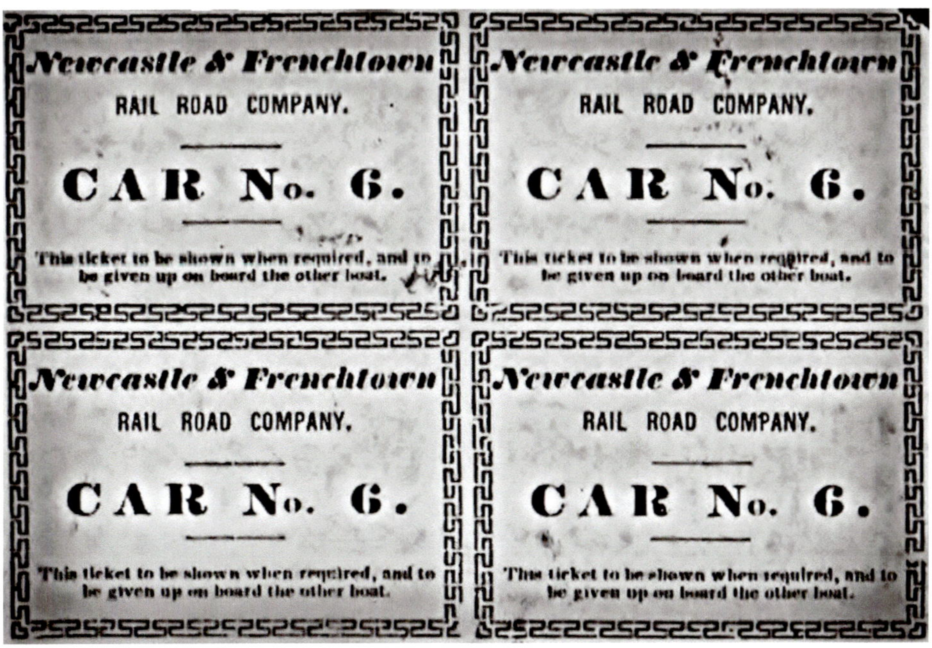

Newcastle & Frenchtown
RAIL ROAD COMPANY.
CAR No. 6.
This ticket to be shown when required, and to be given up on board the other boat.

Newcastle & Frenchtown
RAIL ROAD COMPANY.
CAR No. 6.
This ticket to be shown when required, and to be given up on board the other boat.

Newcastle & Frenchtown
RAIL ROAD COMPANY.
CAR No. 6.
This ticket to be shown when required, and to be given up on board the other boat.

Newcastle & Frenchtown
RAIL ROAD COMPANY.
CAR No. 6.
This ticket to be shown when required, and to be given up on board the other boat.

Travel by this new transportation method was 25 cents per passenger. The fee was just for the train portion, with the entire trip from Philadelphia to Baltimore being three dollars. Despite the high cost for the average American, the rail line saw no shortage of passenger service.

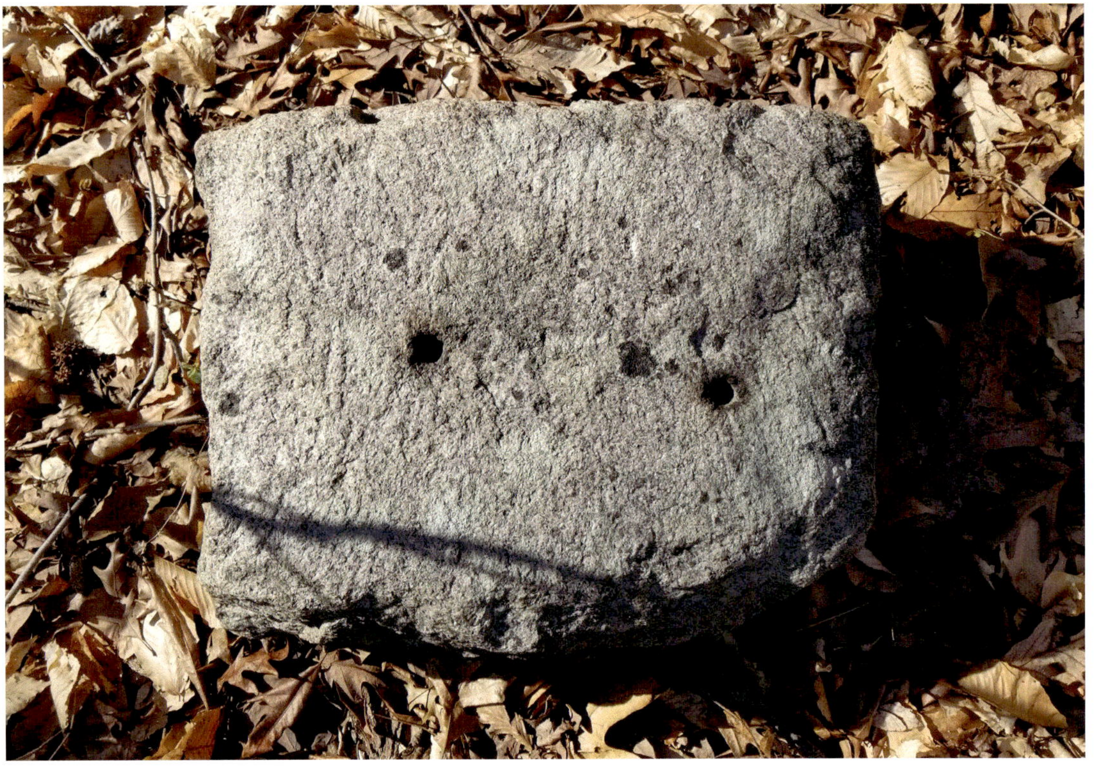

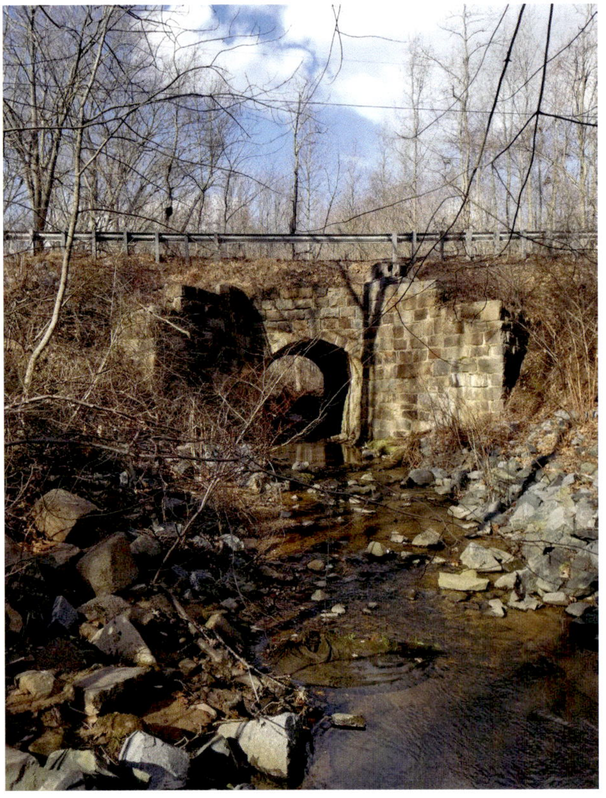

Above: Following the old rail line, you can still find traces of how the tracks were laid. Originally stones were placed, wooden plugs would be used to hold a wooden beam, and then the thin metal rail would be attached. This method proved flawed and the more traditional method of wooden ties would come into being. Almost 200 years later, these stones can still be seen.

Left: Standing today as a silent reminder of this railroad is a stone bridge, located just west of where the line ended in Frenchtown, Maryland. No longer supporting train traffic, today it spans the same creek as it did in 1832, but as a roadway bridge. Standing by the creek one can almost hear the old small steam engine chugging over the rails.

Right: By 1853, the New Castle and Frenchtown had become part of the newly constructed Delaware Railroad. Championed by Judge Samuel Harrington, who would become the railroad's first president, the Delaware Railroad would, by 1856, lay track covering almost the entire state of Delaware, north to south. It now took less than half a day to travel the entire length of the state.

Below: The New Castle station had now moved from the old pier to a new brick structure along the double rail line. Traffic along the entire route was so frequent, rails were laid allowing for trains to travel northbound and southbound simultaneously. Separate freight houses, as the one seen on the left of this photo, were constructed at the busier stations. By the mid-1960s, almost all passenger service had stopped and most of these stations were torn down. [*Photo courtesy of the Delaware Public Archives*]

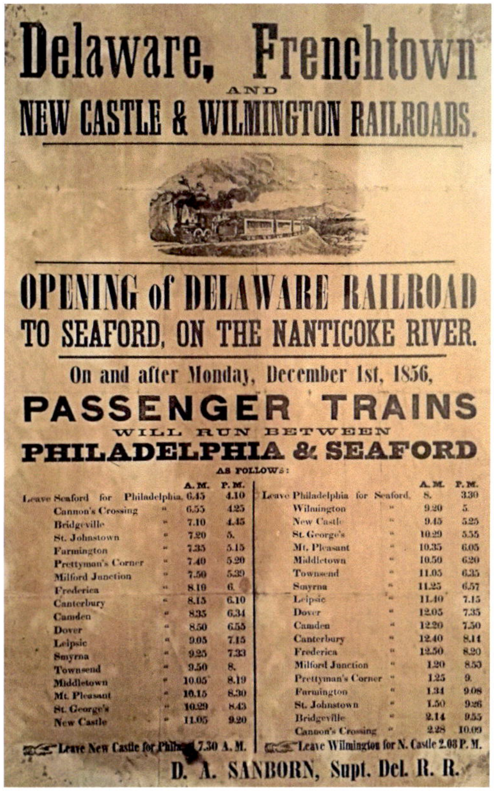

Delaware, Frenchtown
AND
NEW CASTLE & WILMINGTON RAILROADS.

OPENING of DELAWARE RAILROAD
TO SEAFORD, ON THE NANTICOKE RIVER.

On and after Monday, December 1st, 1856,
PASSENGER TRAINS
WILL RUN BETWEEN
PHILADELPHIA & SEAFORD
AS FOLLOWS:

		A. M.	P. M.
Leave Seaford for Philadelphia.		6.45	4.10
Cannon's Crossing	"	6.55	4.25
Bridgeville	"	7.10	4.45
St. Johnstown	"	7.20	5.
Farmington	"	7.35	5.15
Prettyman's Corner	"	7.40	5.20
Milford Junction	"	7.50	5.39
Frederica	"	8.10	6.
Canterbury	"	8.15	6.10
Camden	"	8.35	6.34
Dover	"	8.50	6.55
Leipsic	"	9.05	7.15
Smyrna	"	9.25	7.33
Townsend	"	9.50	8.
Middletown	"	10.05	8.19
Mt. Pleasant	"	10.15	8.30
St. George's	"	10.29	8.43
New Castle	"	11.05	9.20

Leave New Castle for Phila. 7.30 A. M.

		A. M.	P. M.
Leave Philadelphia for Seaford.		8.	3.30
Wilmington	"	9.20	5.
New Castle	"	9.45	5.25
St. George's	"	10.29	5.55
Mt. Pleasant	"	10.35	6.05
Middletown	"	10.50	6.20
Townsend	"	11.05	6.35
Smyrna	"	11.25	6.57
Leipsic	"	11.40	7.15
Dover	"	12.05	7.35
Camden	"	12.20	7.50
Canterbury	"	12.40	8.14
Frederica	"	12.50	8.20
Milford Junction	"	1.20	8.53
Prettyman's Corner	"	1.25	9.
Farmington	"	1.34	9.08
St. Johnstown	"	1.50	9.26
Bridgeville	"	2.14	9.55
Cannon's Crossing	"	2.28	10.09

Leave Wilmington for N. Castle 2.08 P. M.

D. A. SANBORN, Supt. Del. R. R.

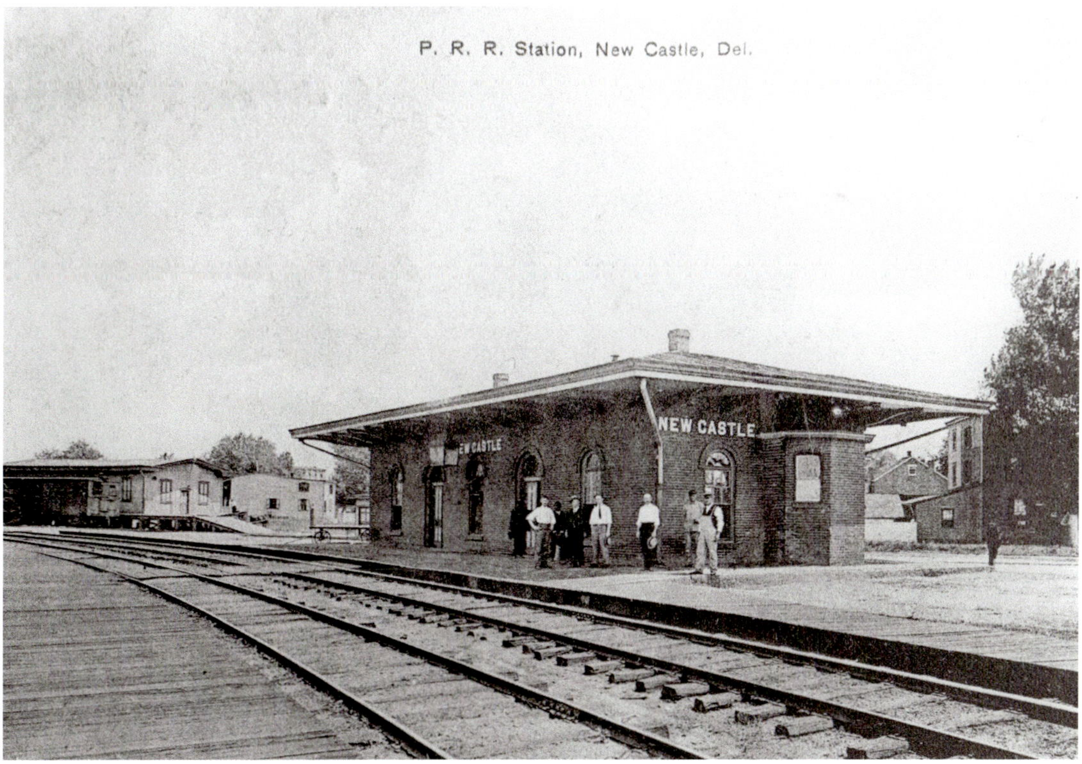

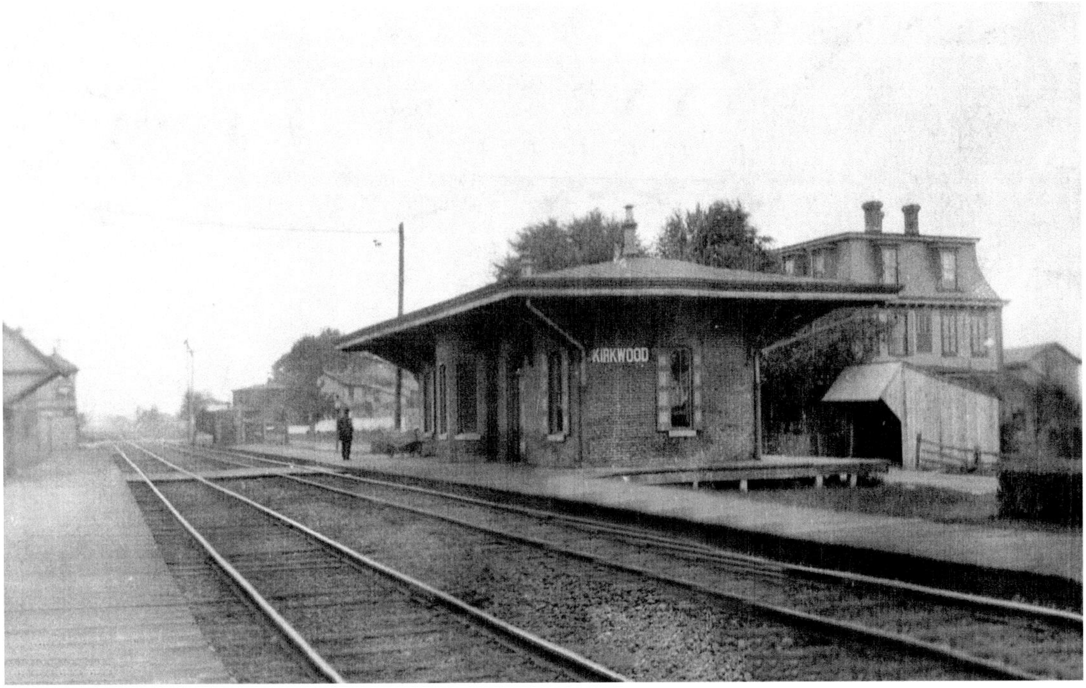

As the demand for access to the new railroad grew, more stops were added, such as this one in Kirkwood, Delaware. Most of the stations were brick and had separate men's and women's waiting areas inside. Smaller stops like Kirkwood were the first to be removed after World War II when passenger traffic began to decline.

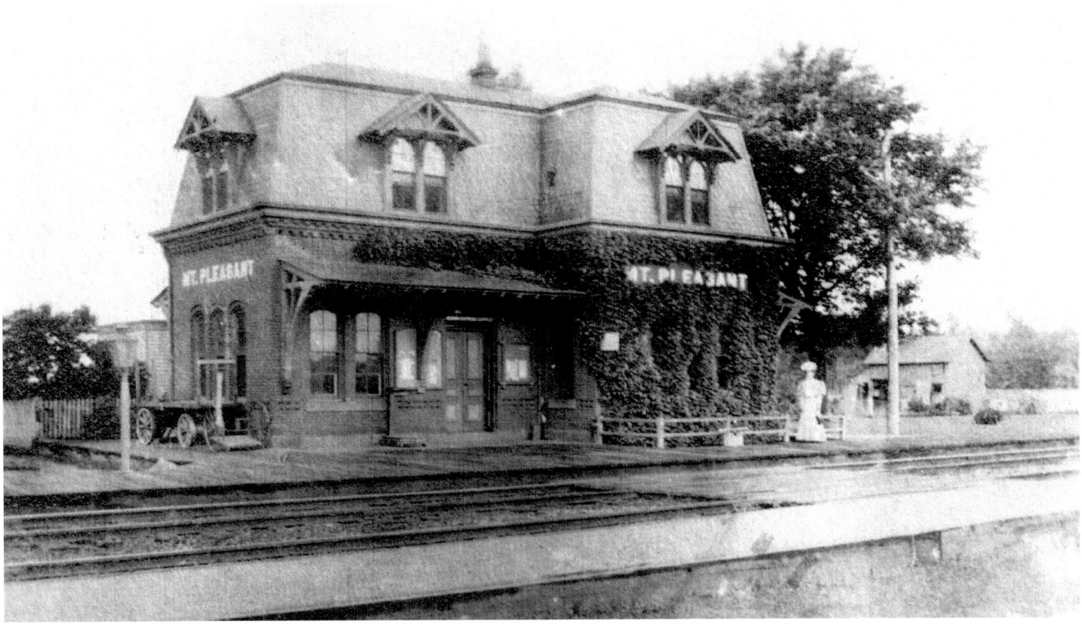

Mt. Pleasant was a major stop for the new Delaware Railroad. A two-story station was constructed, rare for such a rural location at the time. Frequently the train master and his family would reside on the second floor of these stations since housing in rural areas such as this was difficult to find. Mt Pleasant was one of the last passenger stations to see service removed, and sadly it did not survive long after that time.

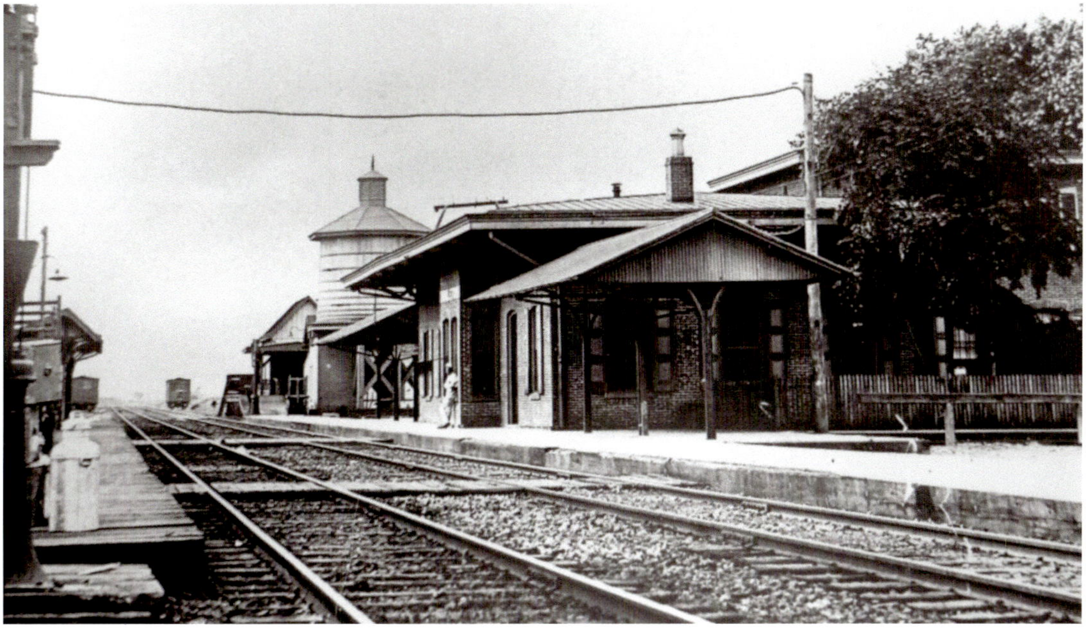

Middletown, Delaware, was a major freight location before the railroad came. Located near some of Delaware's best peach farms, freight traffic was a major reason for a large station to be constructed in the city. Special railroad sidings or yards were constructed directly beside the large canning operations in the area. Today, very little is left of this thriving operation. The two-story station, the freight house, even the double rail, have all been removed. Standing watch now, barely visible above the gravel on the right is a stretch of concrete denoting where these structures once stood. This site is all too common along the areas just talked about in the book. [*Photo courtesy of Jim Bowden & Keith Short*]

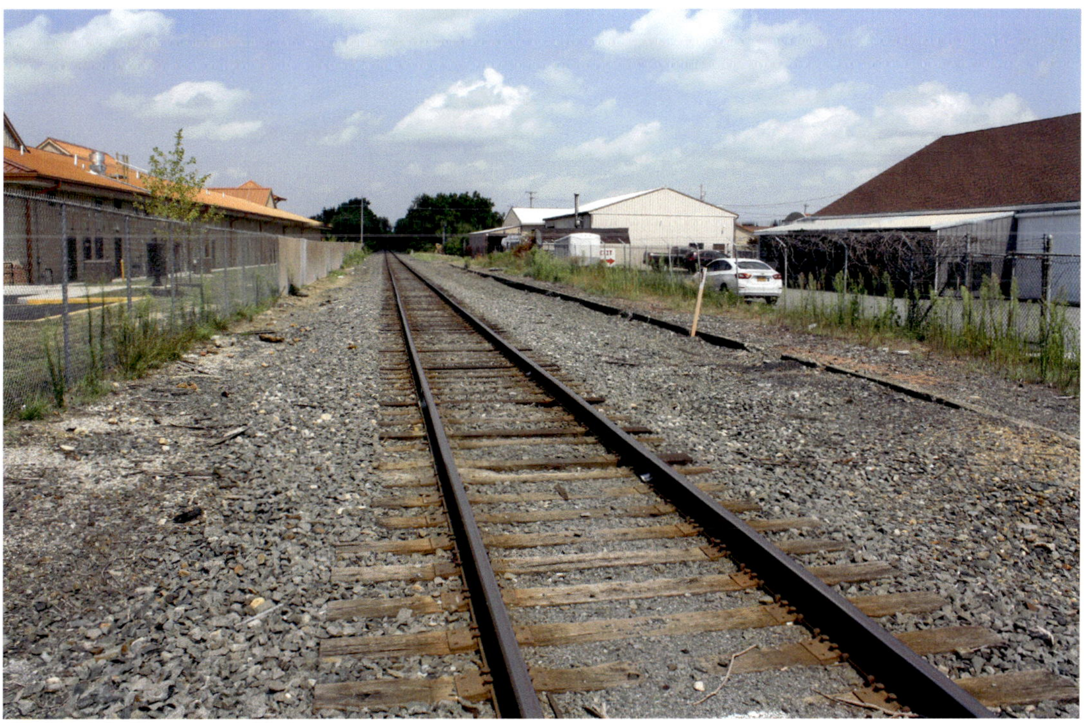

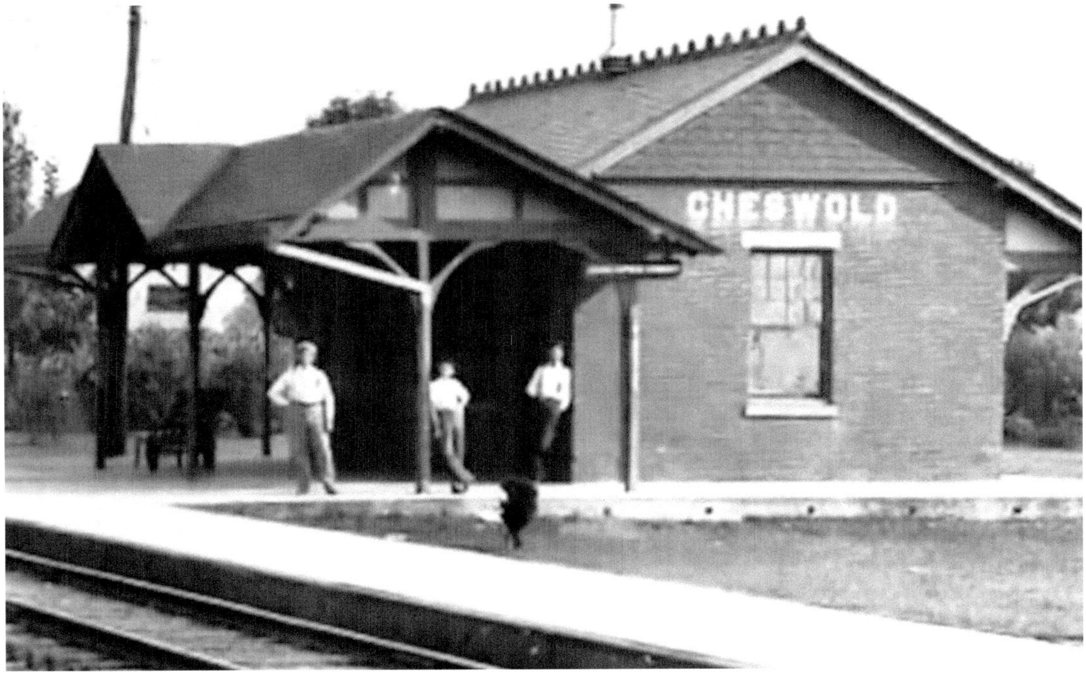

The Delaware Railroad continued its way south into Kent County, with stops large and small. Cheswold grew up around Leipsic Station (as it was first called, because there was no town where the station was built). The name came from the chestnut trees that were planted near the station. [*Photo courtesy of Jim Bowden & Keith Short*]

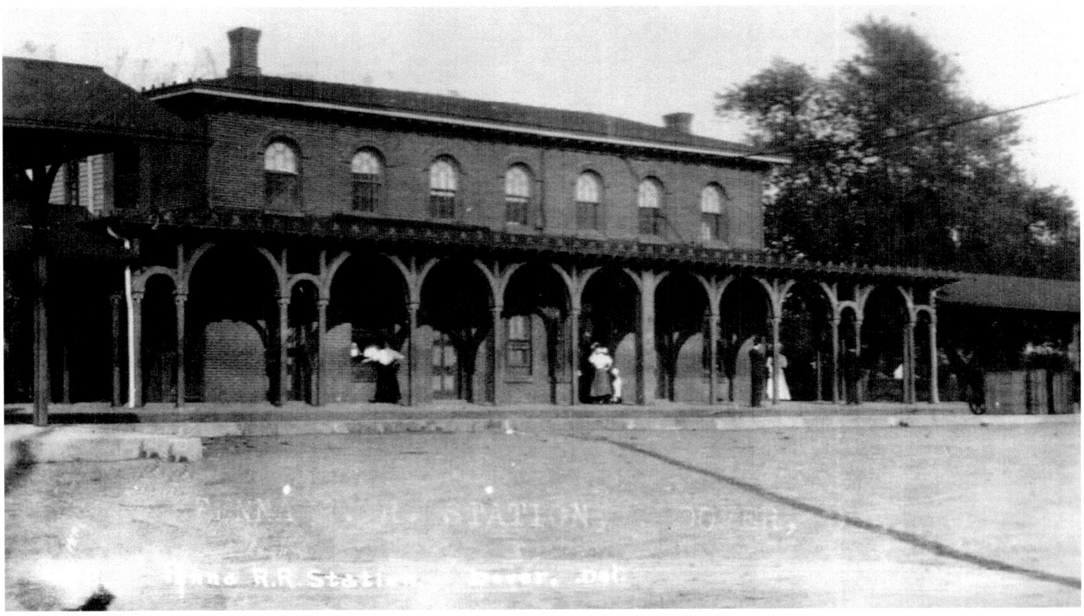

The original railroad station in Dover Delaware was constructed in 1853 with the coming of the railroad. It was the second largest station constructed for the Delaware Railroad at the time. Despite being new, the building was quickly outgrown, and a new station was constructed. Rail traffic was so frequent, an underground tunnel was constructed to allow travelers to safely cross under the rail lines and await their train on the opposite side of the tracks.

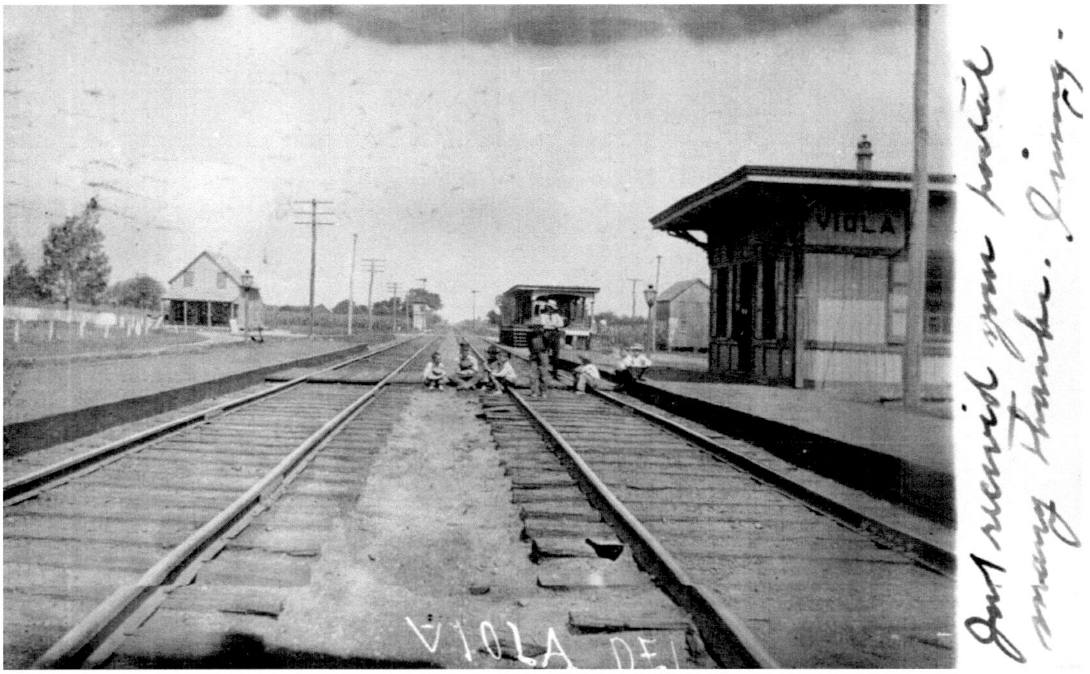

Viola, Delaware, literally sprang up from the countryside that the railroad rumbled through. Called Canterbury Station when the stop was created, it wasn't called Viola until the 1900s when the town was incorporated and named after a local landowner's daughter. Today, as with so many others, just the curbing remains.

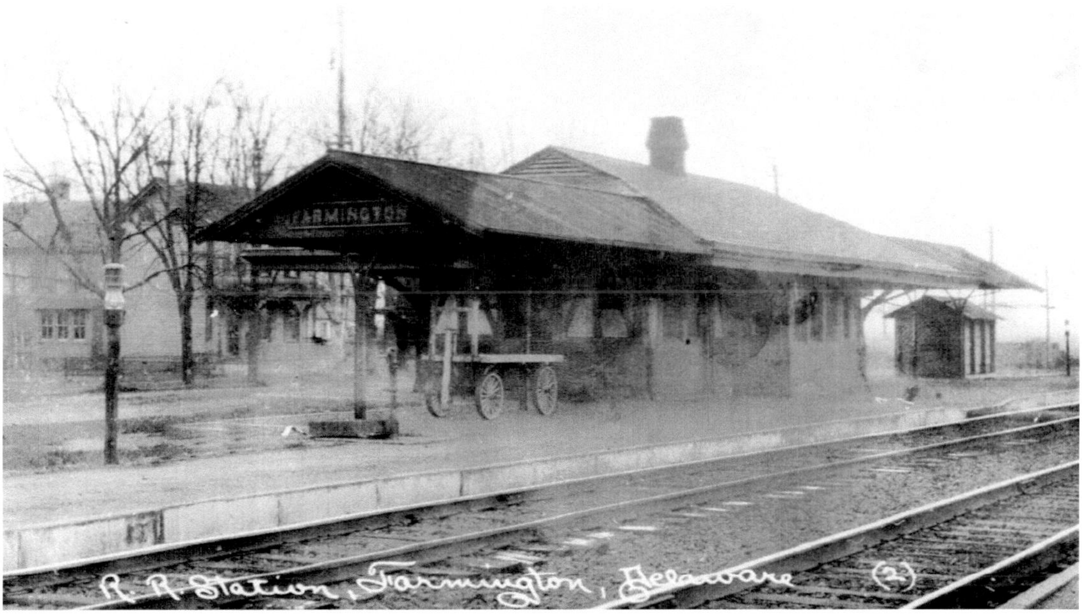

Once one of the busier towns along the railroad, Farmington, Delaware, was already established by the time the railroad came. The town would slowly lose its industry as the city of Harrington became a major yard and interchange for the railroad. The station was removed by the 1940s. [*Photo courtesy of the Farmington Historical Society*]

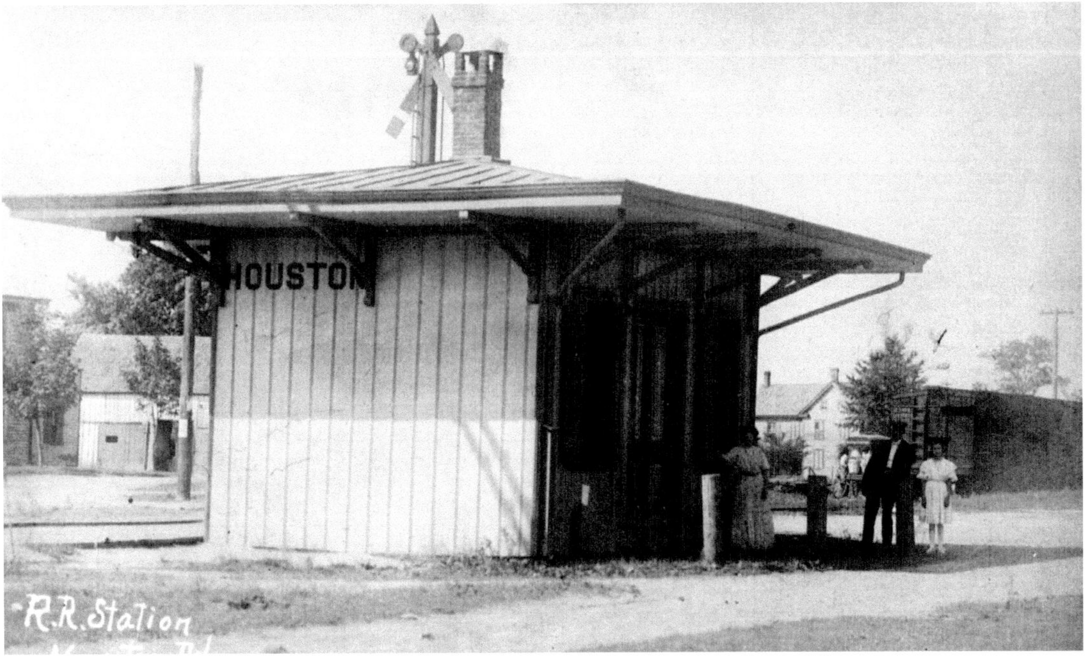

The next railroad that would connect cities and towns was the Junction and Breakwater. Connecting with the Delaware RR at Harrington, its first stop was the small hamlet of Houston, Delaware. While small in size, the town was host to a major canning factory for Libby, McNeil and Libby. While the station was abandoned early on, the cannery maintained a vital freight stop for the railroad for many decades. [*Photo courtesy of Jim Bowden & Keith Short*]

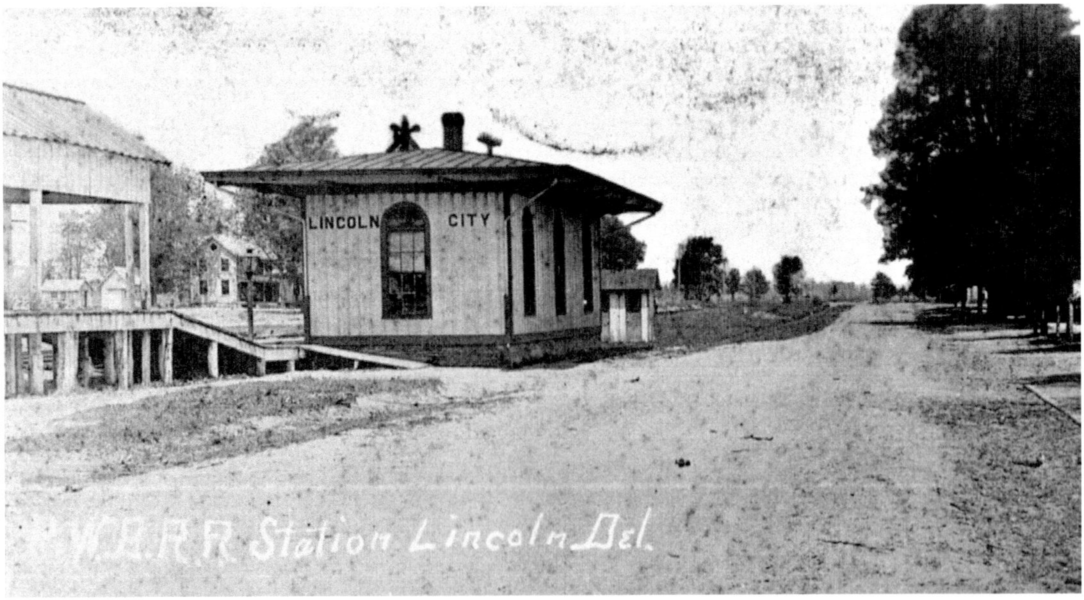

Great plans were the reason for the stop being constructed at Lincoln City, Delaware. Laid out with streets named after U.S. presidents, it was envisioned to become the county seat of government. Since those plans never materialized, the town was never incorporated and both passenger and freight service ended just after the mid-1940s. [*Photo courtesy of Jim Bowden & Keith Short*]

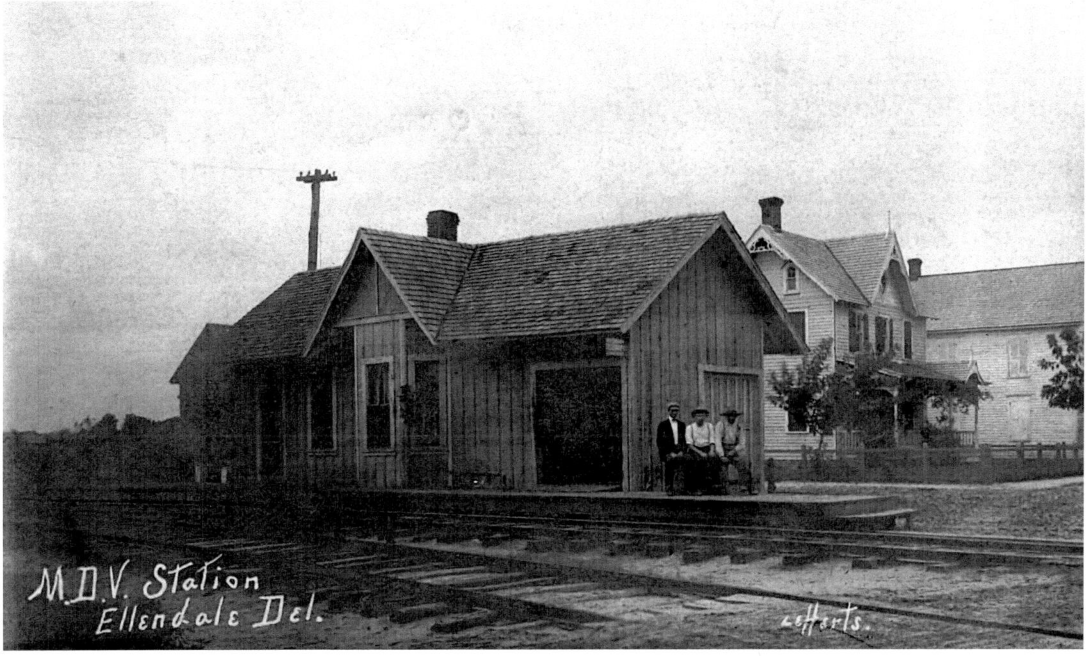

The Junction and Breakwater's next stop was the small city of Ellendale, Delaware. As with so many small towns, when the railroad stop was created, industry came. A sawmill, canneries, basket mills, and even a button factory called this small town home in the early 1900s. [*Photo courtesy of Jim Bowden & Keith Short*]

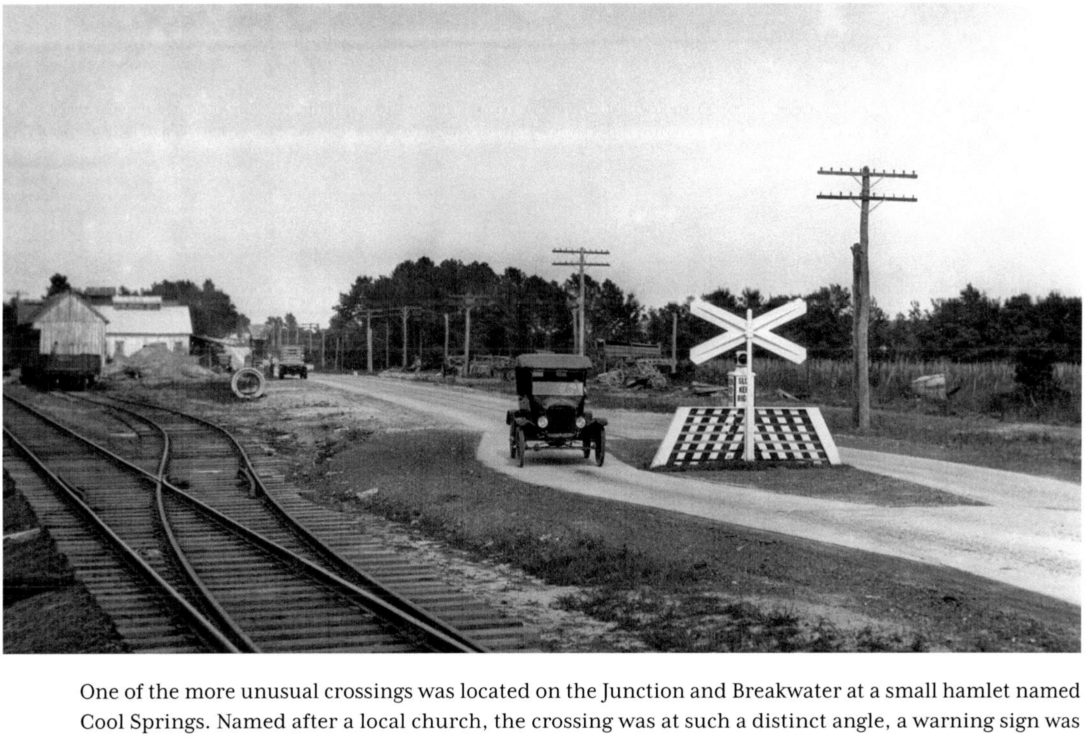

One of the more unusual crossings was located on the Junction and Breakwater at a small hamlet named Cool Springs. Named after a local church, the crossing was at such a distinct angle, a warning sign was placed in the center of the roadway to warn oncoming traffic of the railroad's presence. [*Photo courtesy of the Delaware Public Archives*]

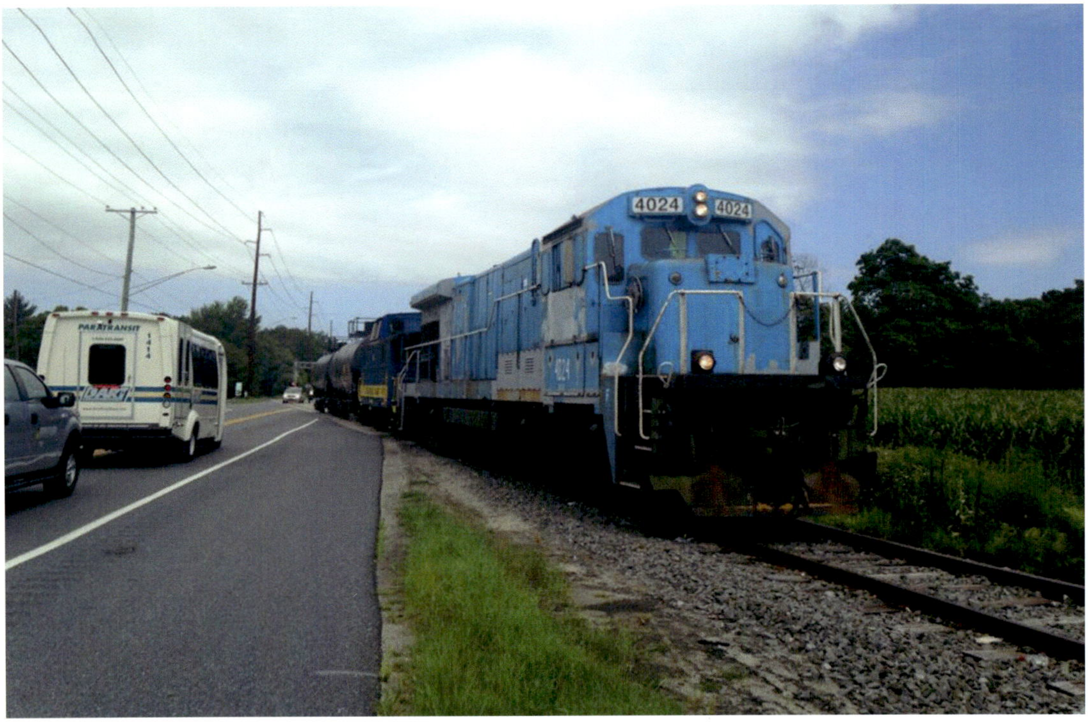

Beginning in 1982, this section of the railway was operated by Delaware Coast Line Railway. By now the unique crossing had overhead lighting, but it still maintained its odd configuration. Most of the rail traffic traveling beyond this point was now chemical cars being taken to SPI Pharma in Lewes, Delaware. Due to major maintenance issues, the line was abandoned and removed in 2018, ending the last remnants of any rail traffic to the beaches in Delaware. [*Photo courtesy of Jim Bowden & Keith Short*]

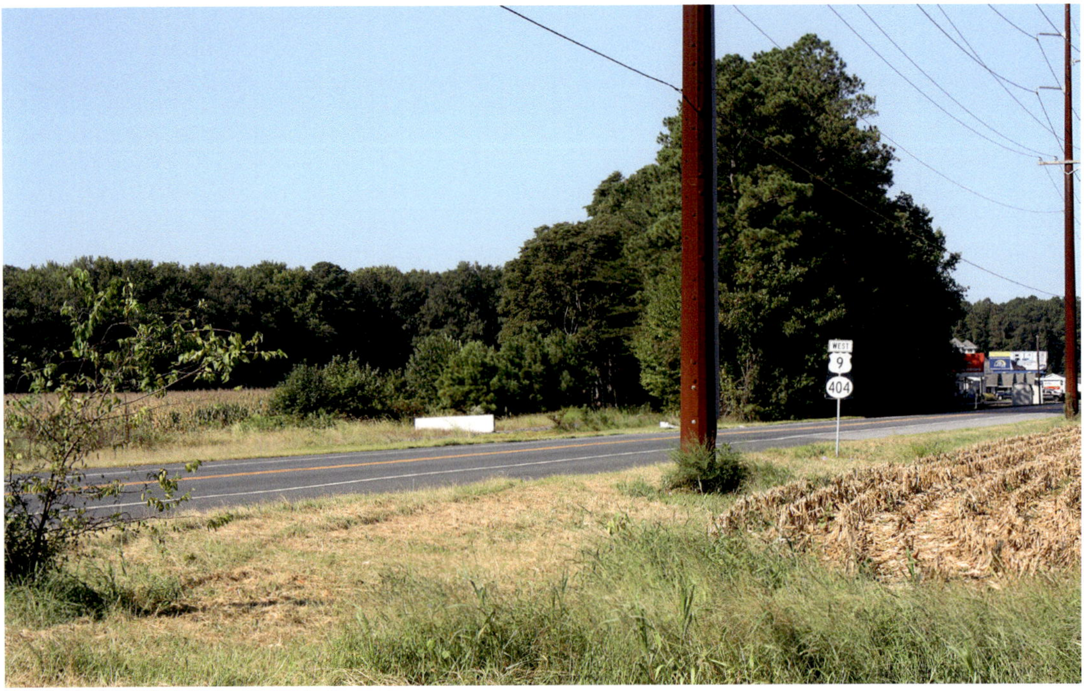

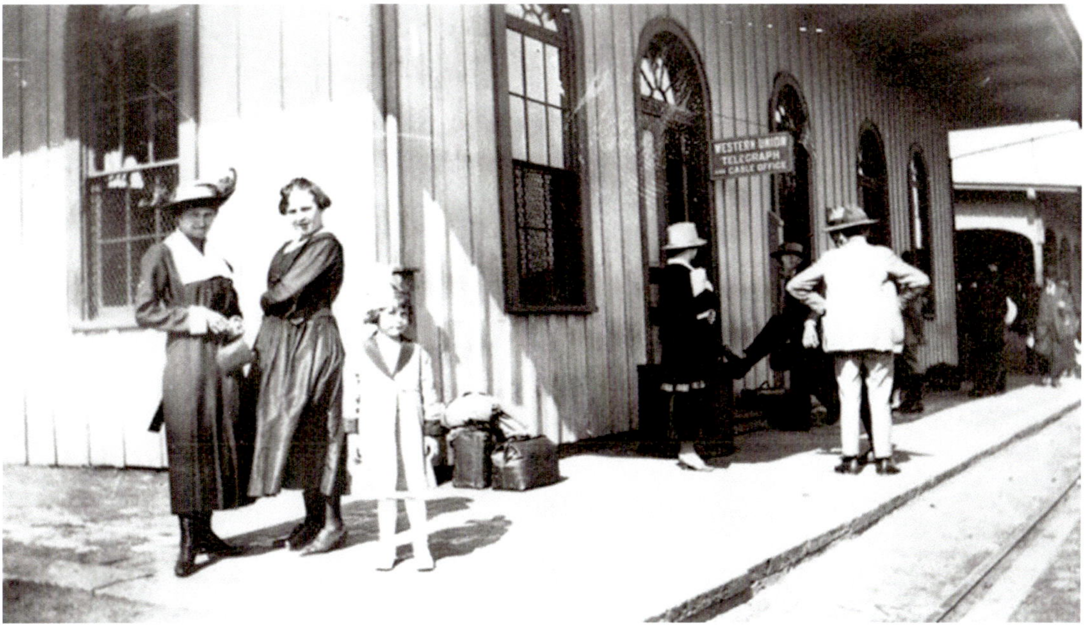

Lewes, Delaware, was once the end of the line for the Junction and Breakwater Railroad. Today a tourist town along the beaches of Delaware, in its heyday, seafood packing was king. By 1938, the Consolidated Fisheries Company was the largest seafood plant in the United States. While tourism played a role for the train, freight was the major reason for the railroad coming to the area.

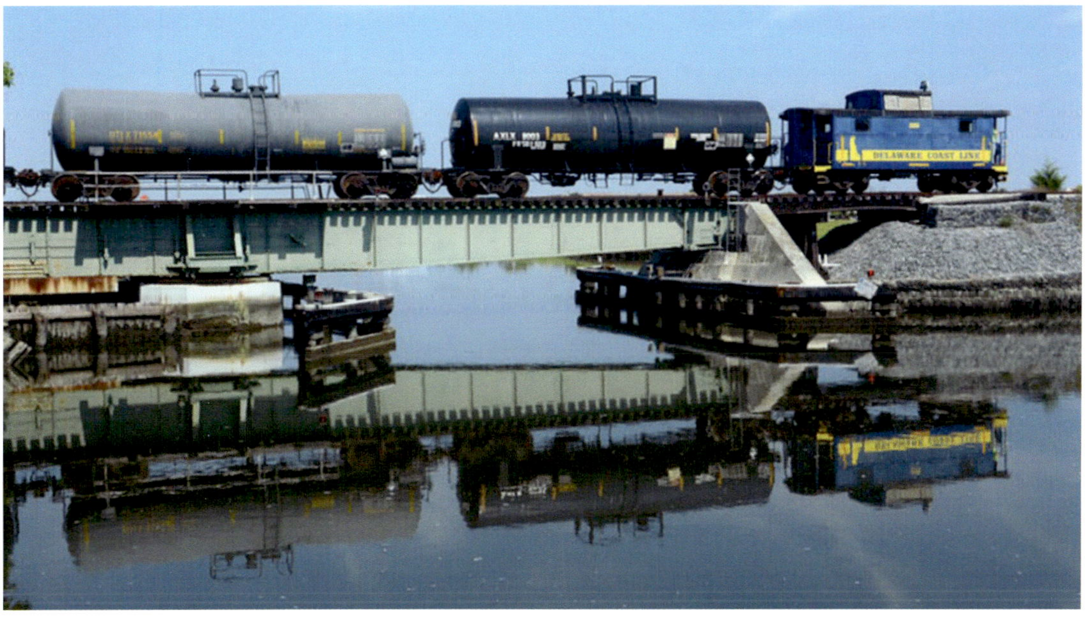

Located on the south side of Lewes, the swing bridge was a local favorite site for many. The bridge was frequently photographed over the decades. The bridge allowed the train to reach the fisheries and later the military post that was located at Fort Miles.

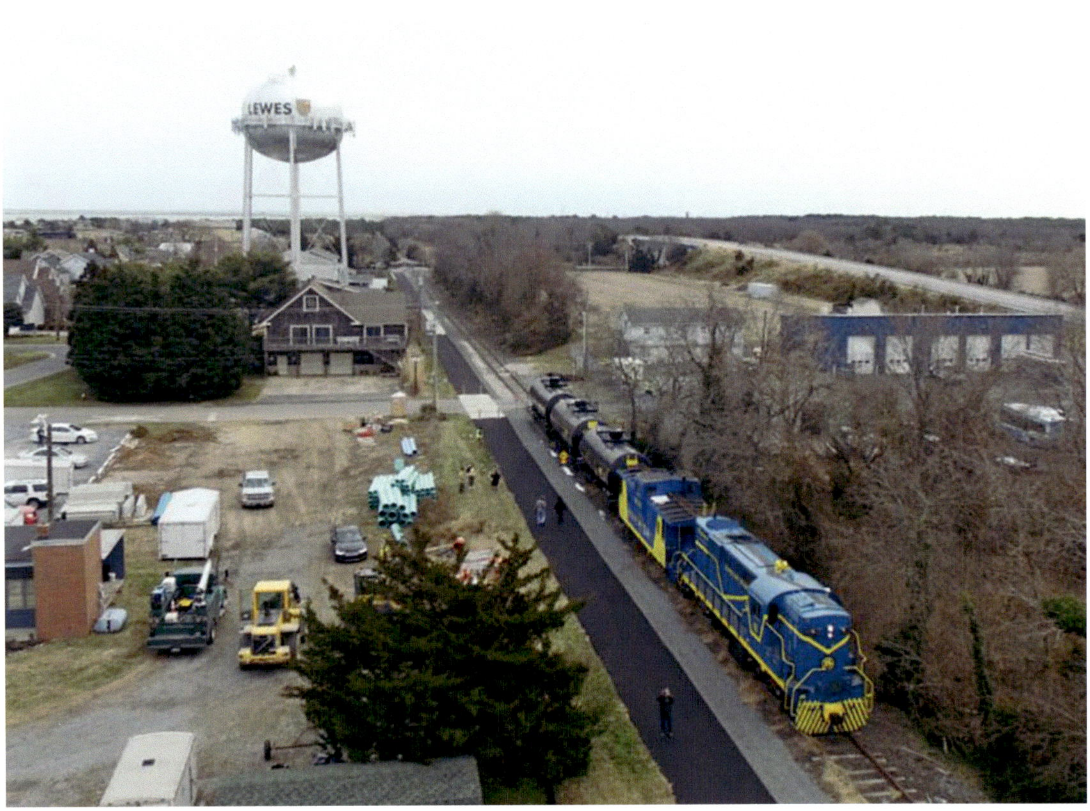

It was this bridge's demise that ended service on the railroad in the area. Found to have dropped over six inches, it was quickly placed out of service. With the cost too high to repair or replace the bridge, it was decommissioned. The last train left Lewes, Delaware, on November 6, 2017, ending nearly 150 years of rail service to the area. Today, most of the rail bed has been convert to the Junction and Breakwater Bike Trail, and the swing bridge is being moved to an area near the library.

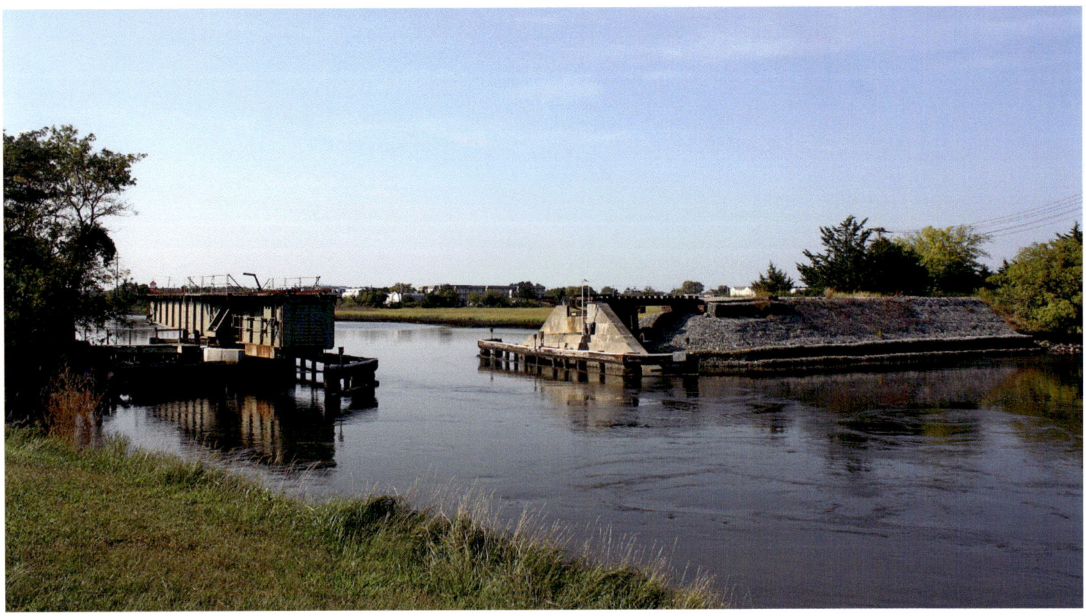

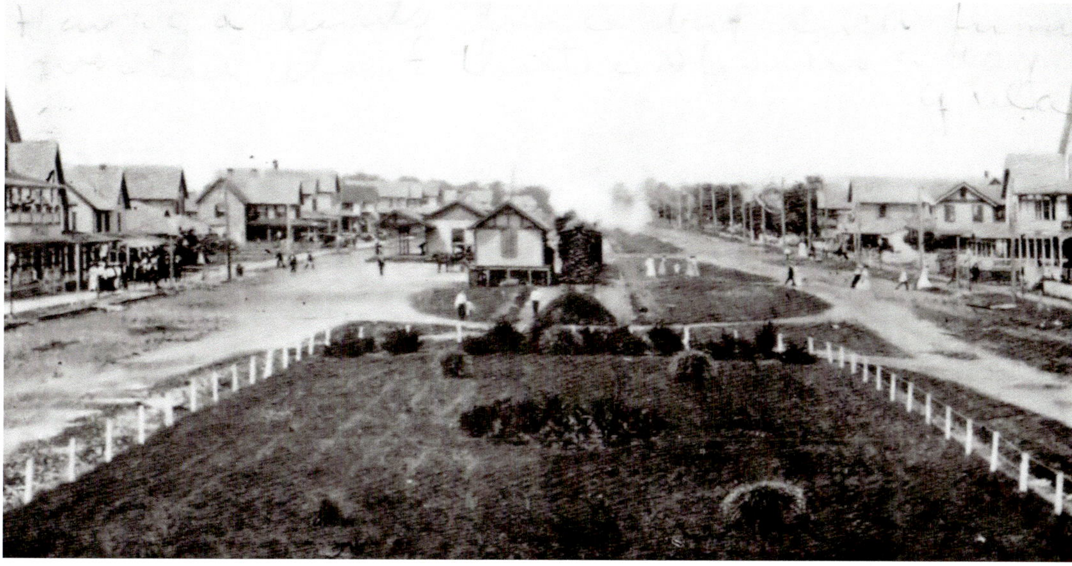

Rehoboth, Delaware, became the Junction and Breakwater's final stop in 1878. Constructed right down the middle of Rehoboth Avenue, the train brought passengers to the growing seaside resort town. Passengers could now board the train in Philadelphia for a weekend getaway to the beaches. But without freight service to enhance revenues, passenger service ended in 1928. Now one of America's most popular beach resorts, there is no sign of the trains that once filled the street.

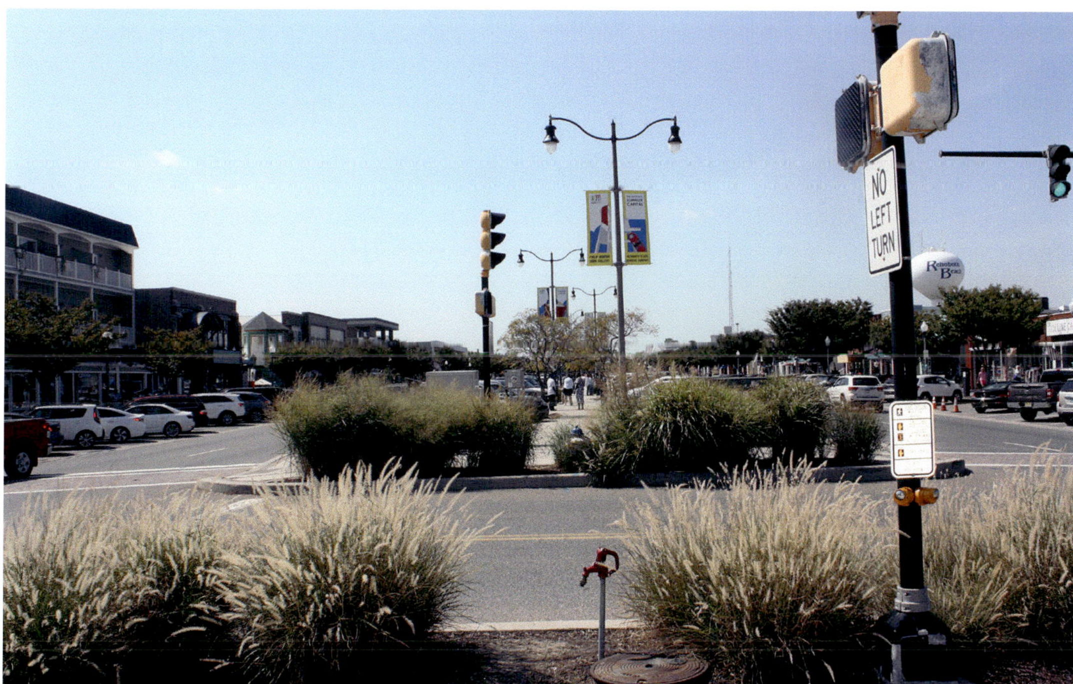

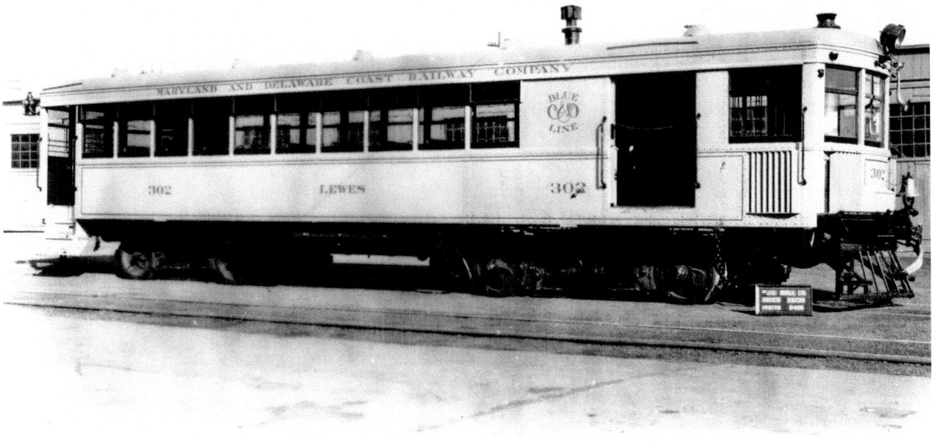

For a brief period, the resort towns kept their passenger train service when the Maryland and Delaware Coast Line Railroad took over service from Denton, Maryland, to Lewes, Delaware. Running this small self-propelled passenger car, they could reduce cost. The car was destroyed by fire in 1931 and the freight business on the line came to an end one year later. The rails were removed between the 1960s and as recently as 2018.

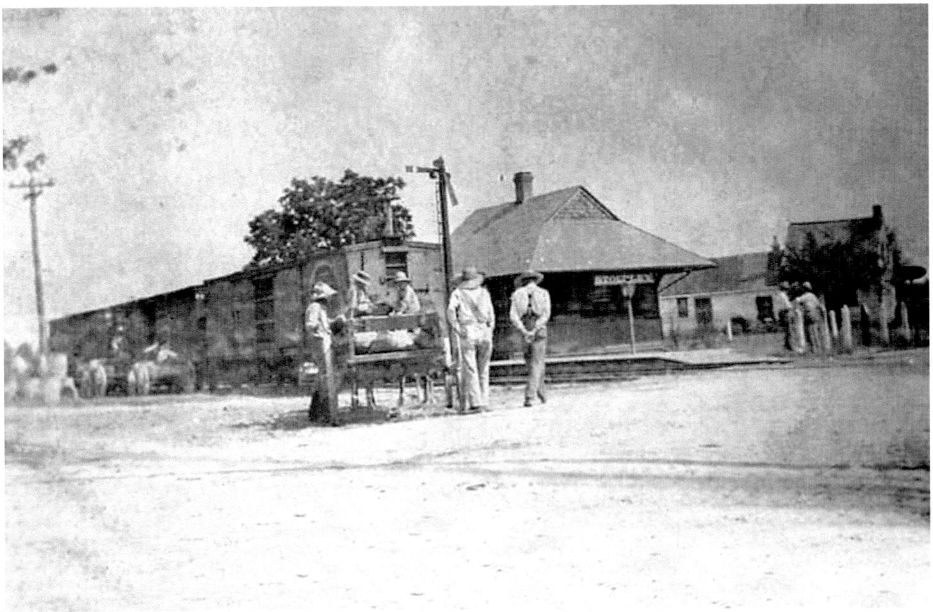

The Breakwater and Frankford was the next railroad to connect rural parts of Delaware to the major cities of the northeast. Strawberries, tomatoes, and other produce reigned supreme in the fields of the area. Stockley was the first stop south of the junction at Georgetown. There was nothing more than a few homes, and farms would come from miles around to ship their goods by rail. [*Photo courtesy of Jim Bowden & Keith Short*]

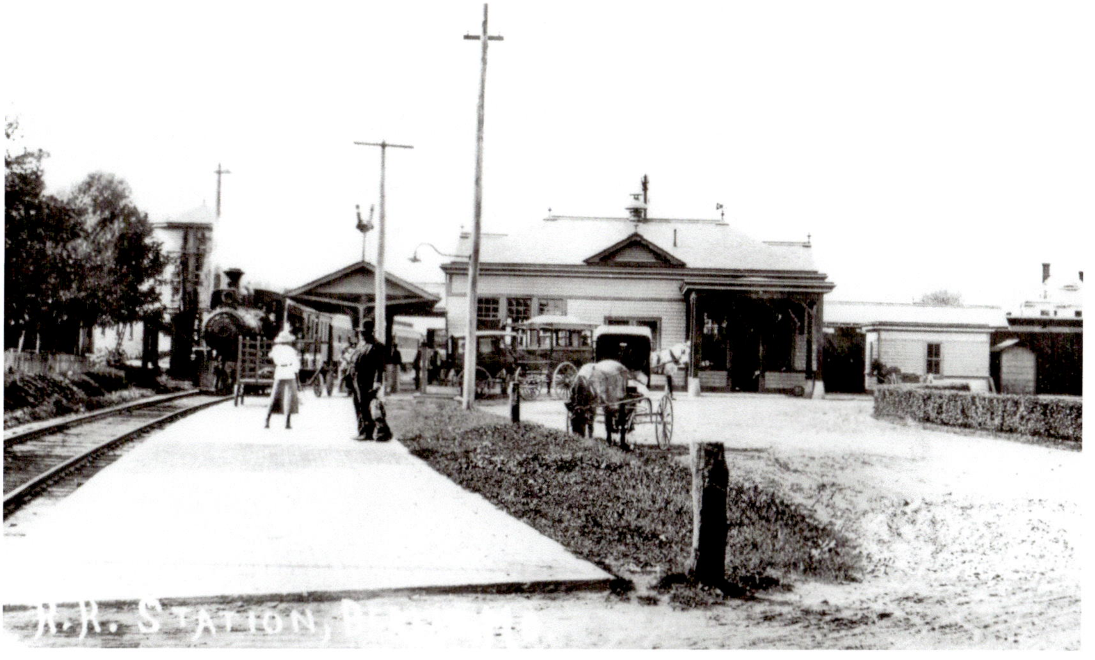

Berlin, Maryland, originally had two passenger stations as both the Breakwater and Frankford, as well as the Wicomico and Pocomoke Railroad, also came through in 1868. The two rail lines decided a Union Station would serve them better. Passenger service ended in the 1930s and freight service in the 1950s. But recently, the freight service has returned. The original freight station now resides on property of the Berlin Fire Department and is used for storage.

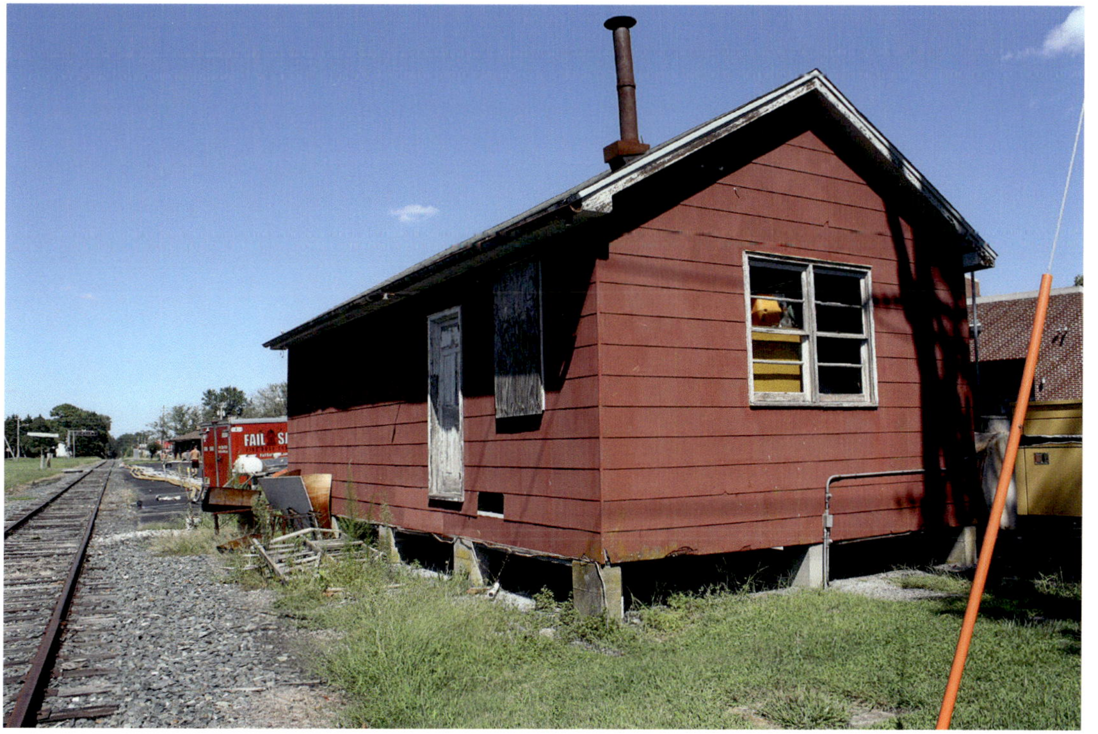

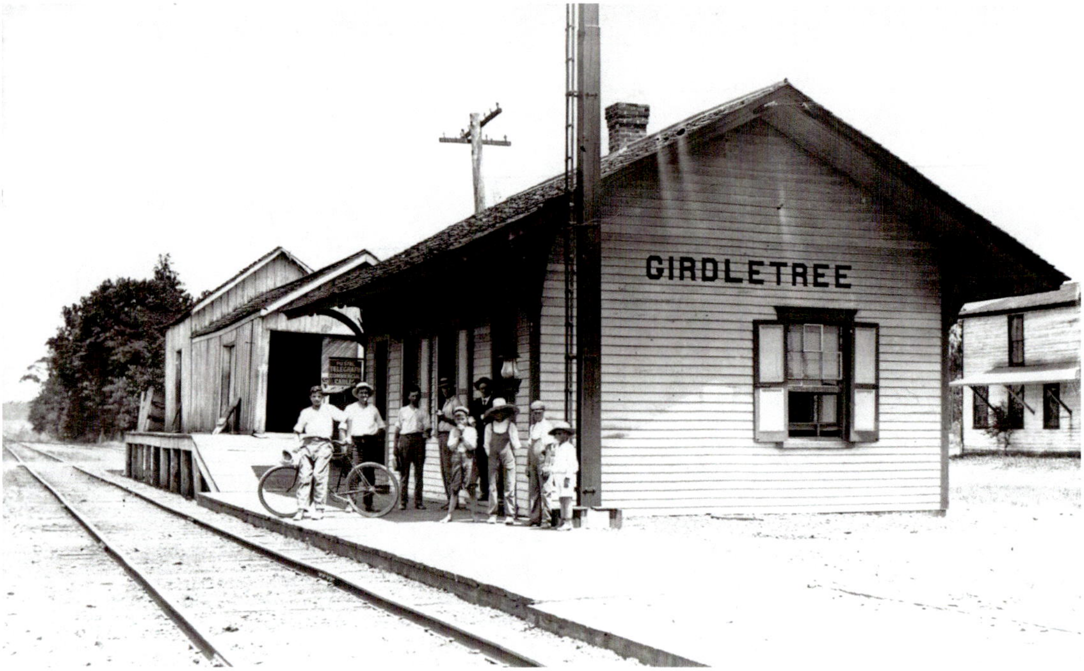

Some towns along the railroad grew up after the rails came while others such as Girdletree Maryland moved to the railroad when it came through the area. A fishing community when it began, the inhabitants moved inland when the railroad placed a stop and freight house at their location. As with so many of these small communities, once freight demands dropped, the railroad ceased operations. The original station and freight house exist, with the station now property of the town. The end of the line was Franklin City Virginia which provided access to the bountiful seafood harvest around Chincoteague Virginia. [*Bottom photo courtesy of Jim Bowden & Keith Short*]

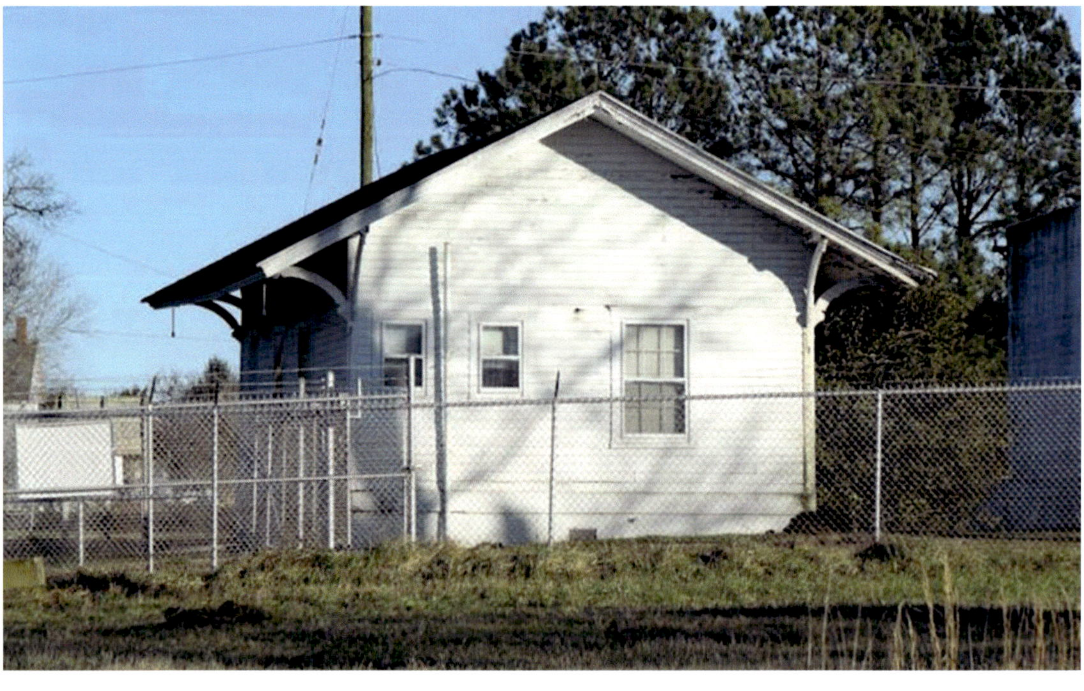

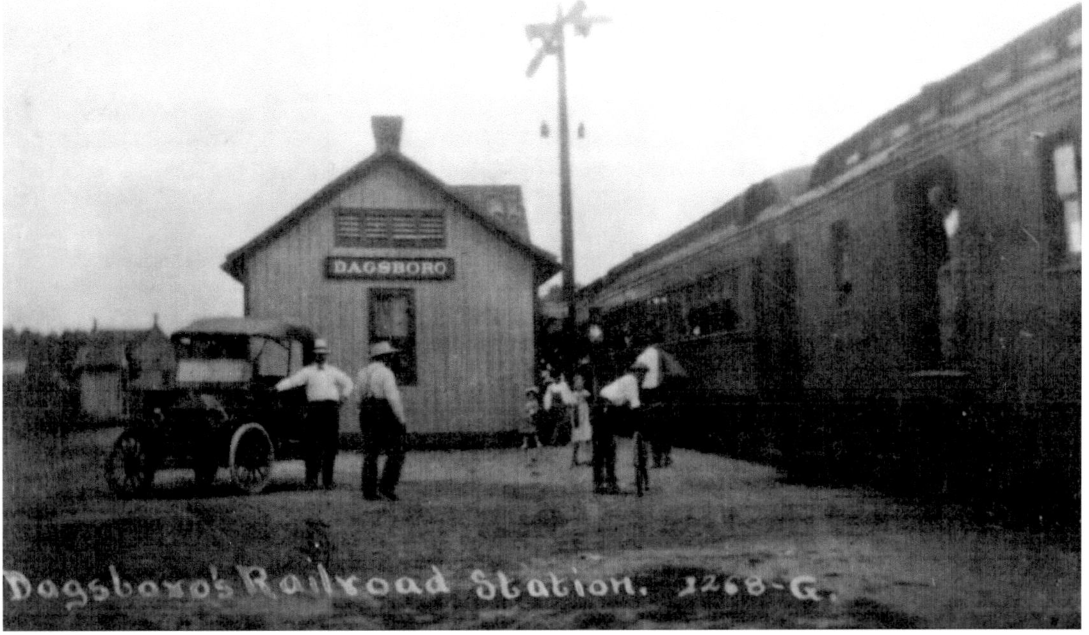

Originally known as Dagsborough, this small town was a fishing village originally, but over time became another well-known produce freight stop along the rail line. Most of these small towns averaged populations under a thousand, but the produce values easily topped hundreds of thousands of dollars a year in the early 1900s.

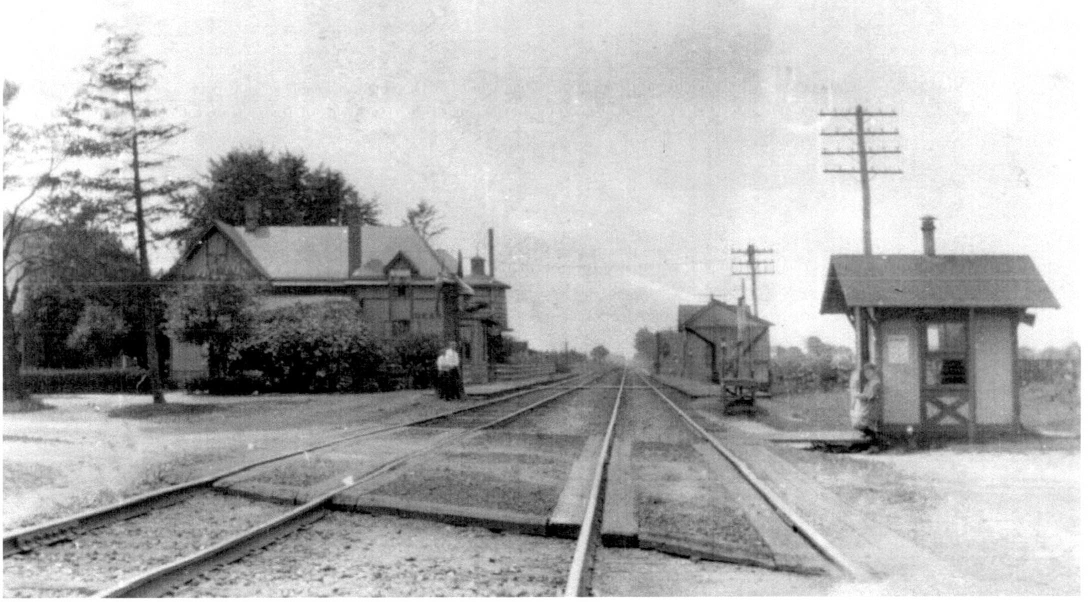

As the railroad grew in power, more and more stops were added to the lines crisscrossing the Delmarva Peninsula. One such stop took the name of the building that was destroyed to make way for the iron horse. Bear Delaware was little more than corn fields and cattle farms when the railroad came through. A local tavern, legend says, laid in the path of the railroad's construction, so it was demolished. Called Bear Tavern, the name stuck. Today, Bear is home to the Amtrak maintenance facility for its coaches.

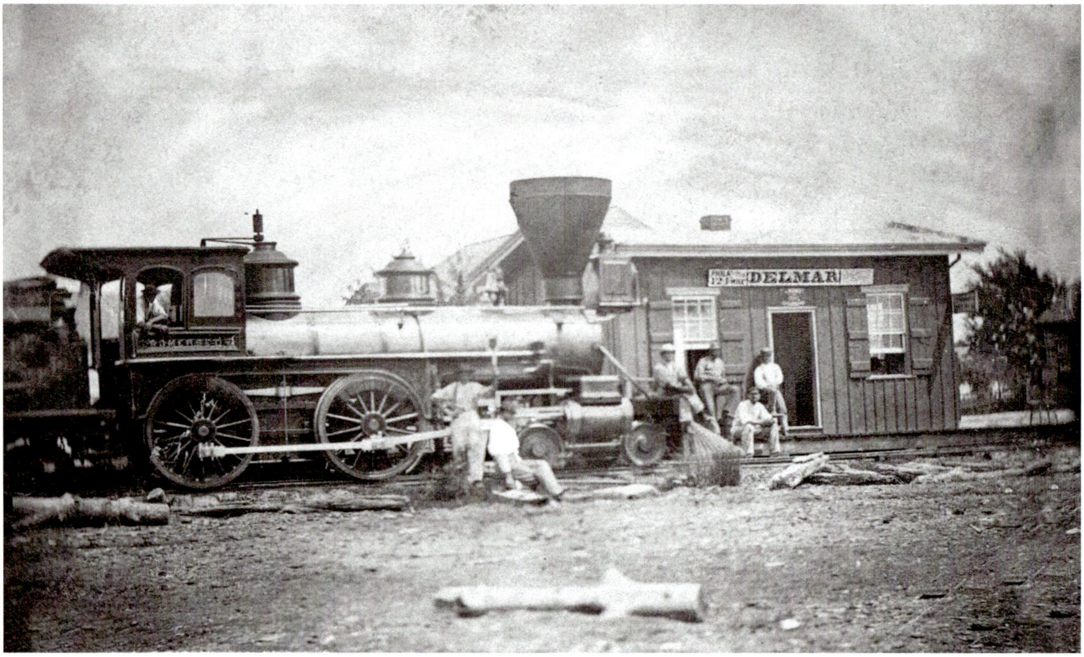

Railroads were granted their right-of-way for the rail beds by each state. So, the Delaware Railroad had to stop at the Delaware-Maryland state line. This created another rail town, Delmar, Delaware, and actually Delmar, Maryland, too. The first station was little more than a shed as the town was the end of the line. The Maryland Railroad had a charter to the other side of the line, so by creating a town that straddles the state line, they found a loophole to permit rail traffic to proceed north and south. The town quickly grew, as did the rail depot here. The railyards were established in 1859, and the junction became a beehive of activity. [*Photo courtesy of Jim Bowden & Keith Short*]

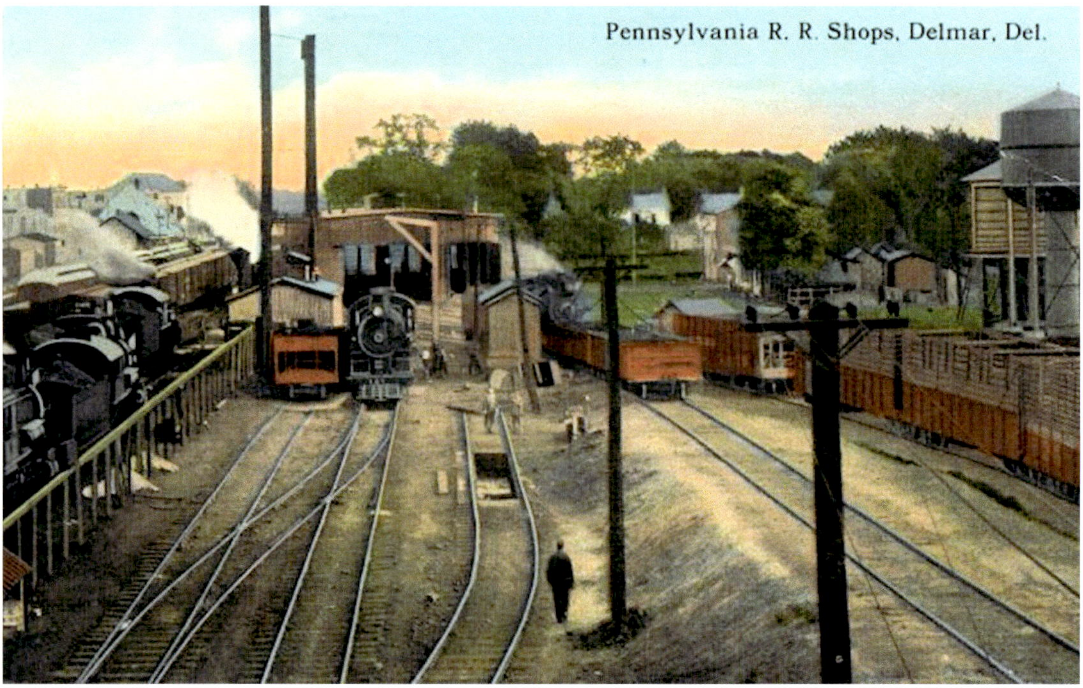

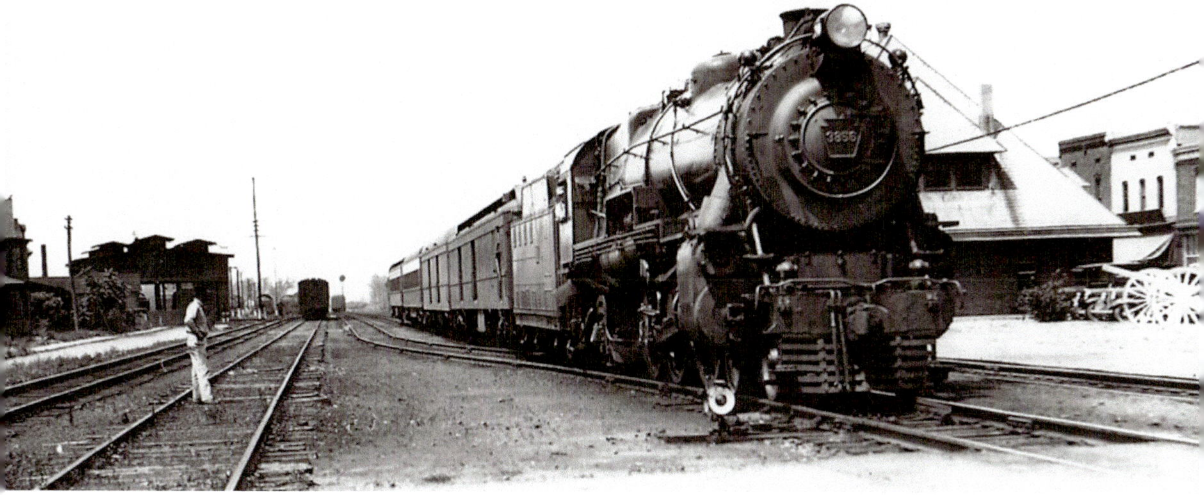

Following the construction of the New York, Philadelphia, and Norfolk, the rail hub became even larger and more important. Hundreds of passengers and thousands of tons of freight would pass daily through this vital railroad junction. Following the consolidation with the Pennsylvania Railroad, Delmar became a regional control point of all the rail traffic in the area. One of the last passenger stations to lose traffic in 1965, today only a caboose museum and a highball stand watch over a single track.

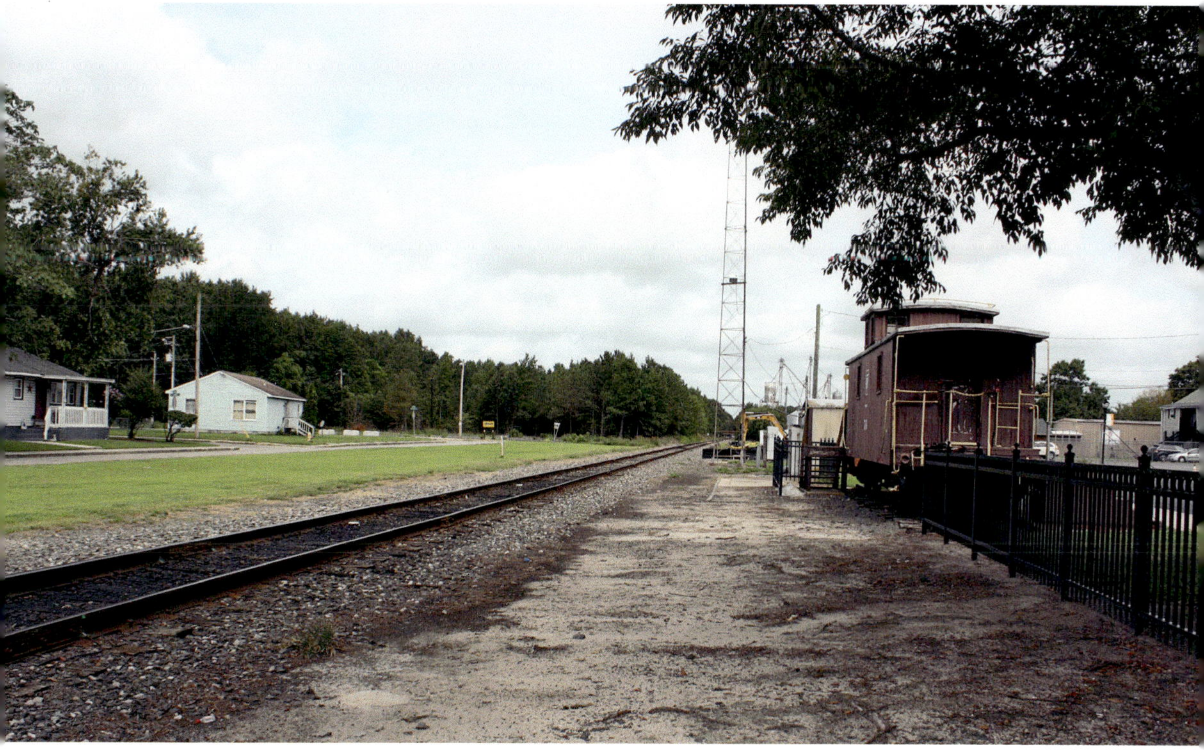

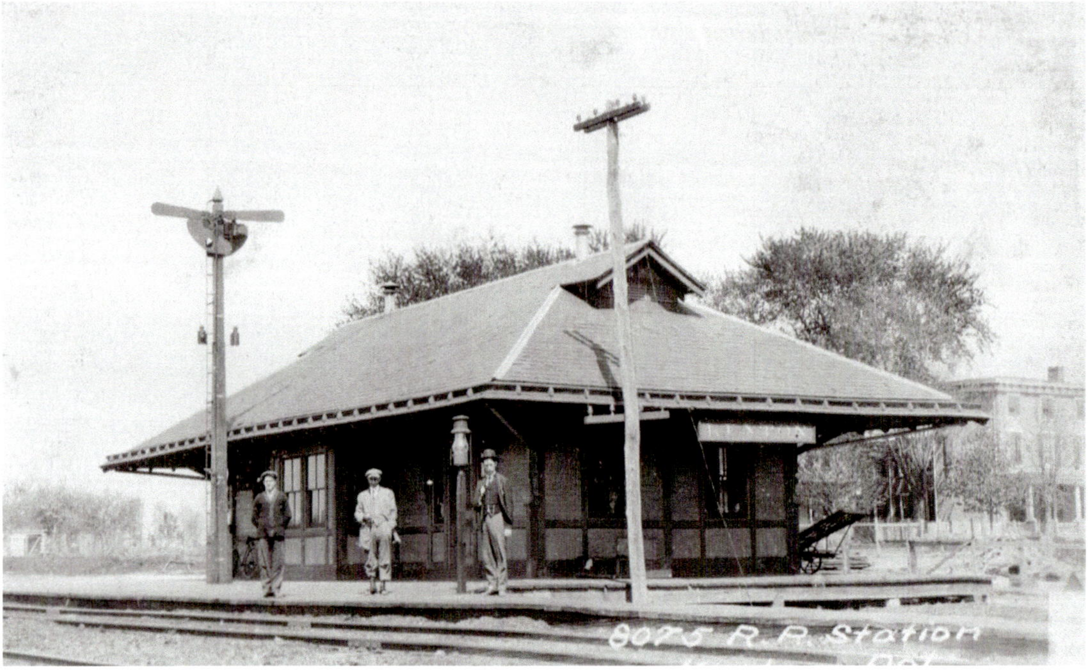

While the center of the area now had train service, there were plenty of agricultural areas that still needed access to this vital transportation link. The Maryland & Delaware Railroad began in 1854, connecting several areas to the Delaware Railroad. Communities such as Kenton, Delaware, were given stops along this new line. The crossroads community grew little and the railroad operations ceased in the mid-twentieth century. Little remains today but memories of what was and a few physical sights of long forgotten rails in the area.

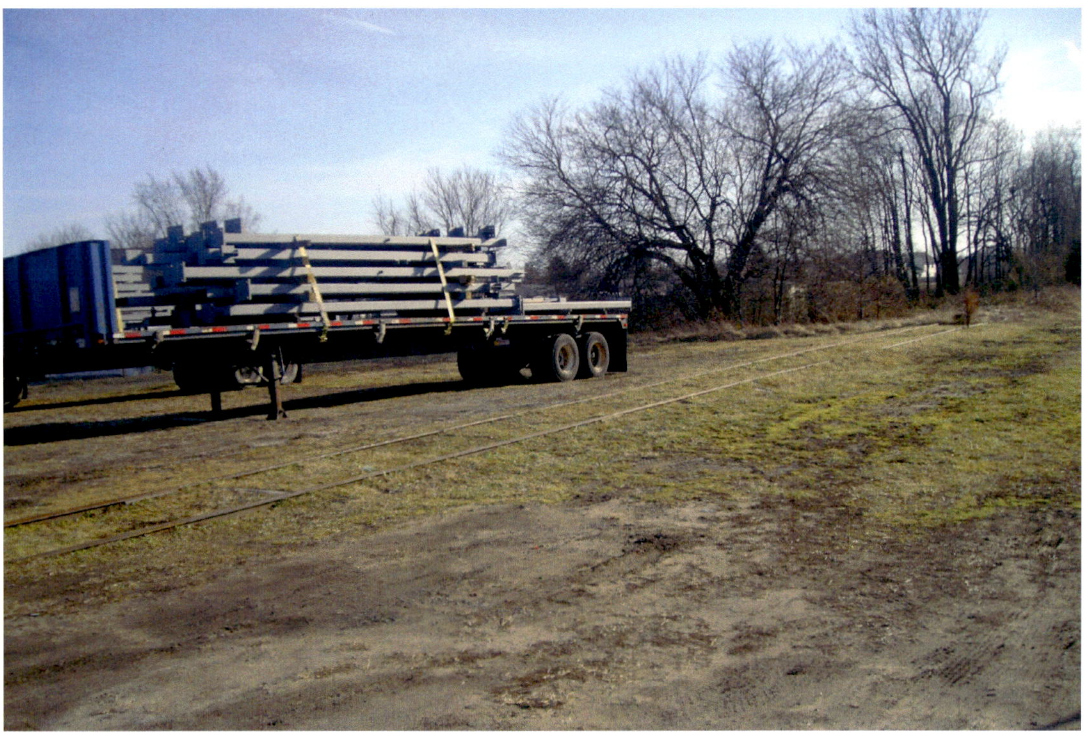

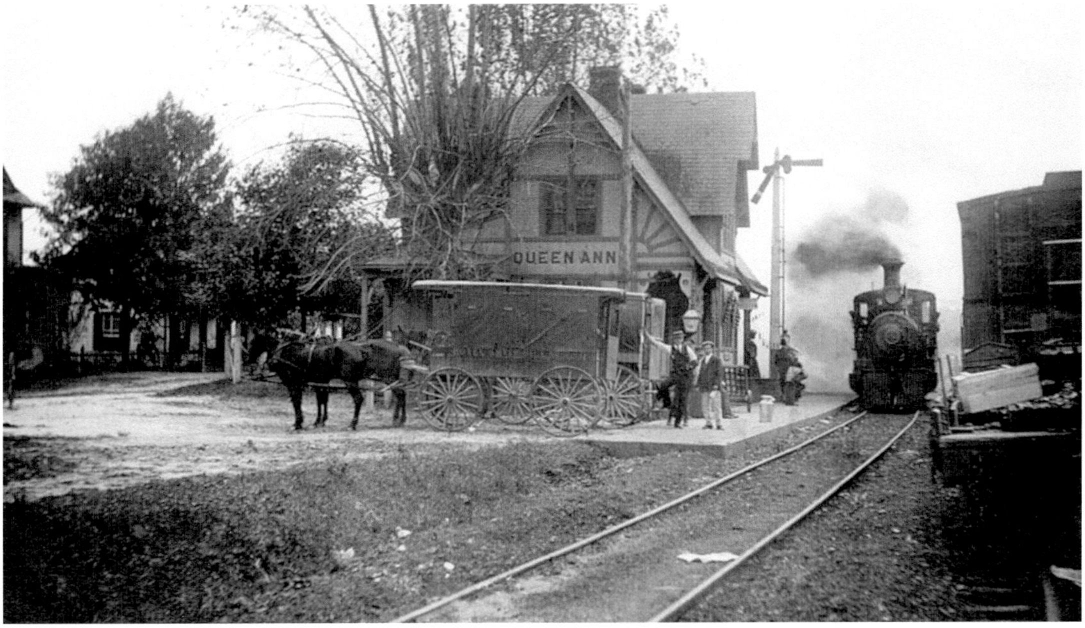

Queen Anne, Maryland, owes its very existence and name to the railroad. Arriving in 1882, the railroad chose to place a stop along the creek that ran through the area. Purchasing land from Jacob Morgan, the area's primary resident, a Victorian-style station was constructed near his home and designed to match the architecture. The railroad then gave the station the name of Queen Anne, named after the county where the newly built station was placed. [*Photo courtesy of Jim Bowden & Keith Short*]

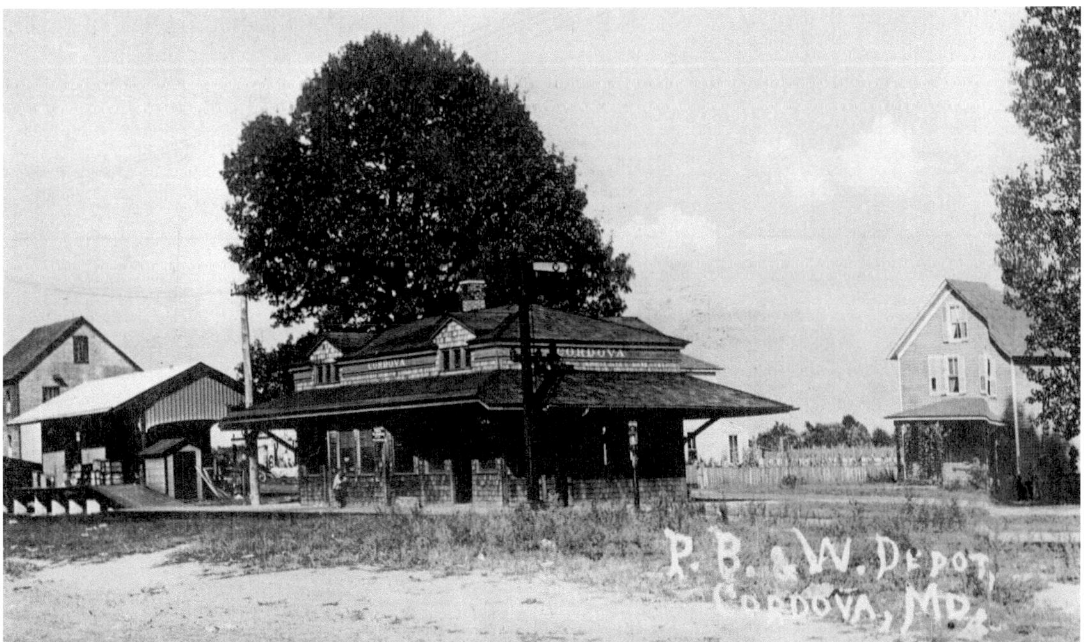

Cordova, Maryland, was another of these rural stops along the Maryland & Delaware Railroad. Founded along a small creek, the area is also believed to be the birthplace of Frederick Douglass. While passenger service ended in the 1930s, freight service along this line continued into the 1970s.

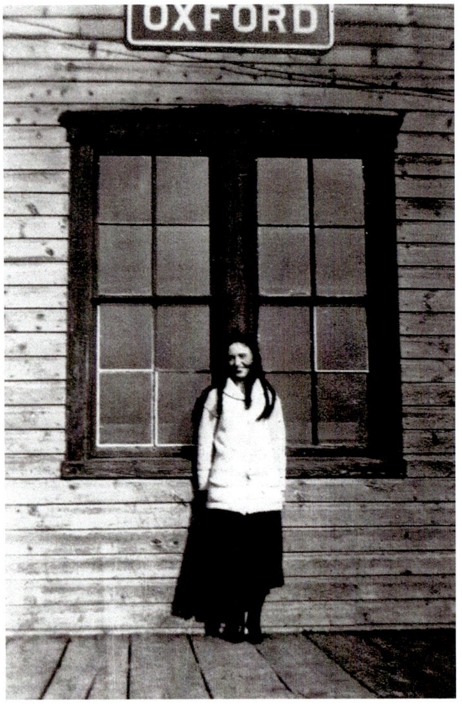

The location for the end of the line for the Maryland & Delaware was Oxford, Maryland. By the time the railroad came the city in 1871, the area was already a busy harbor and seafood harvesting area. Steamboats from across the bay brought travelers while the oyster industry grew rapidly. By the beginning of the 1900s, the oyster beds began to reach the end of their harvesting and slowly the rail traffic began to die off. Today, only a small outbuilding remains to serve as a reminder of the area's rich past. [*Photo courtesy of the Oxford Historical Museum*]

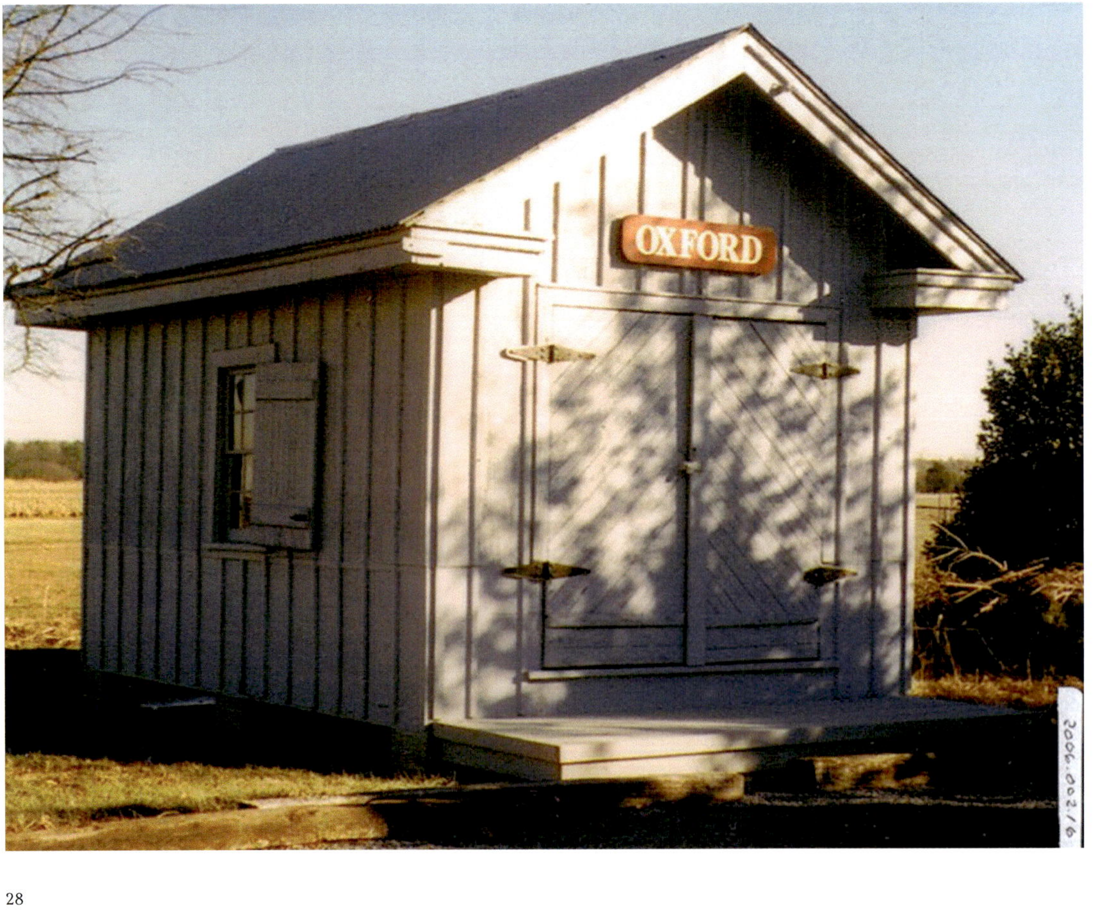

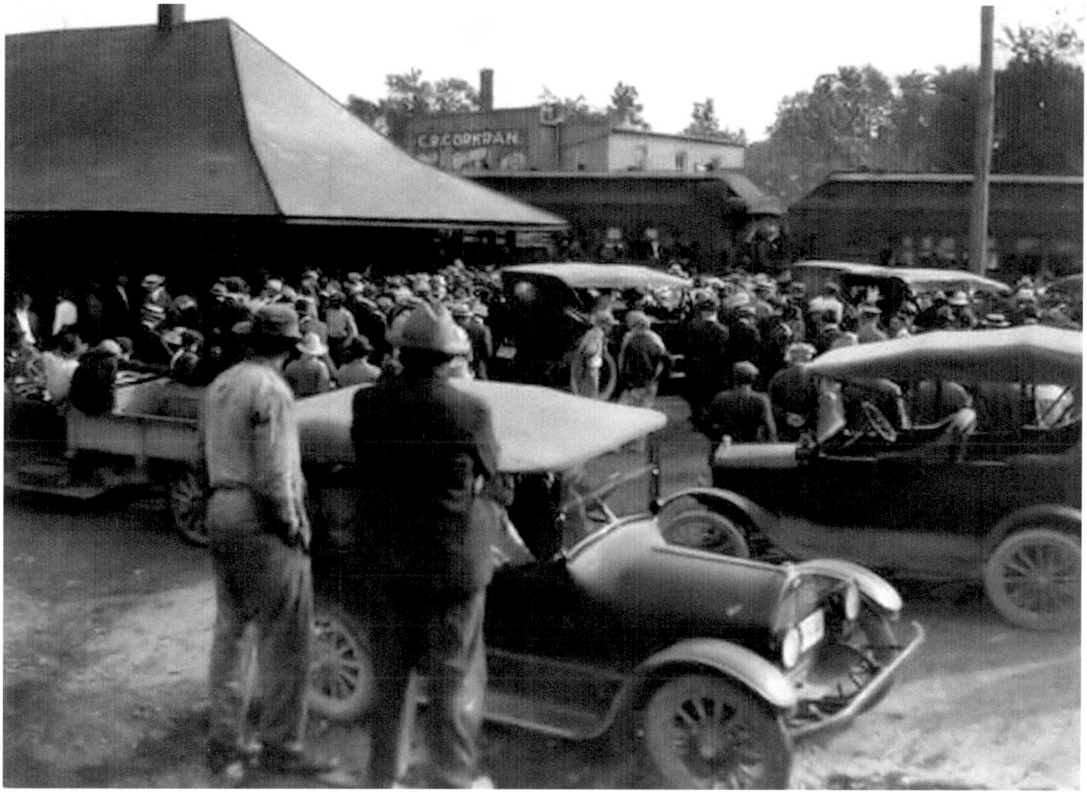

In 1866, the Dorchester and Delaware Railroad was constructed, linking the busy port of Cambridge, Maryland, to the Delaware Railroad at Seaford Delaware. At the time of the railroad's arrival, Cambridge was the largest town along Maryland Eastern Shore. The passenger station was one of the busiest, especially during the World War I and World War II eras. Seafood and vegetable canning was the main industry of the area. Phillips Packing house won defense contracts during both wars. At their peak, Phillips employed over 10,000 workers and shipped boxcars of freight daily to be sent overseas to feed our troops. *[Top photo: courtesy of Jim Bowden & Keith Short. Right photo: courtesy of the Dorchester County Historical Society]*

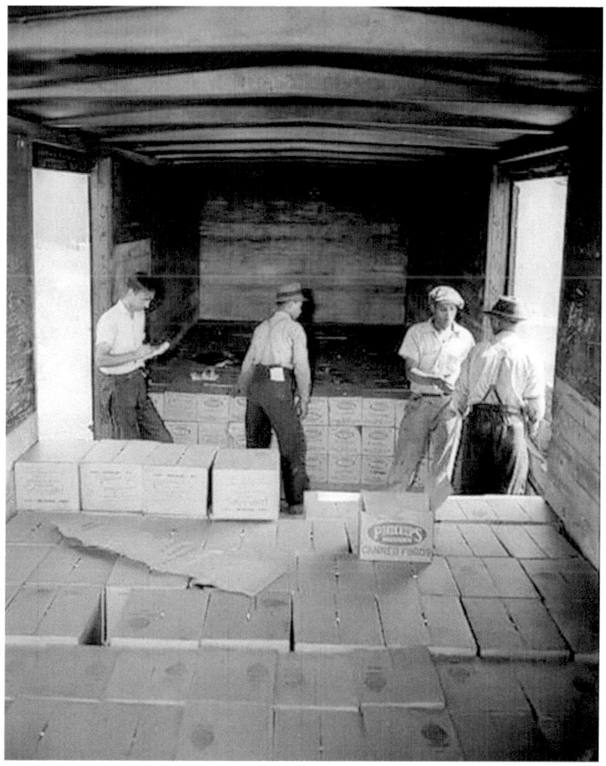

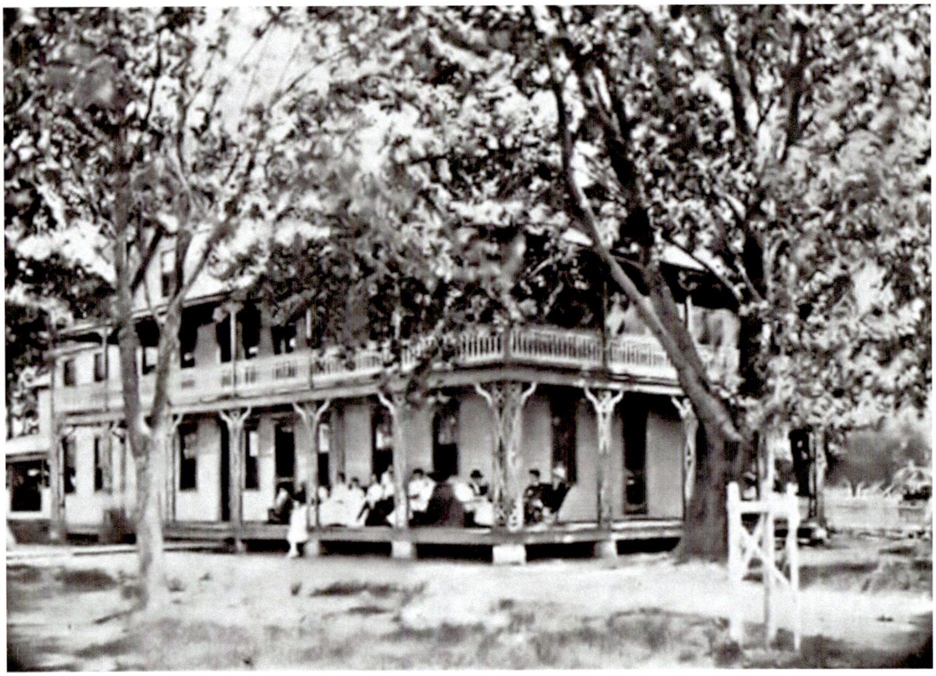

1873 saw the construction of another railroad traveling east to west, this time across the upper half of the peninsula. The Smyrna and Delaware Bay Railroad was a bold project, with the goal of connecting two ferry terminals, one at Woodland Beach, Delaware, and the other was originally to be located at Rock Hall, Maryland. The Maryland portion of this line had begun in 1868, under the name Kent County Railroad Company. When the two lines merged, the western end was then changed to Tolchester Beach, Maryland. Woodland Beach, Delaware, was at the time of the railway's construction, one of the most popular seaside resort in the Mid-Atlantic area. A car float service was constructed to move people and freight traffic to New Jersey. The entire area was devastated by a hurricane, and by 1899 the railroad had been abandoned.

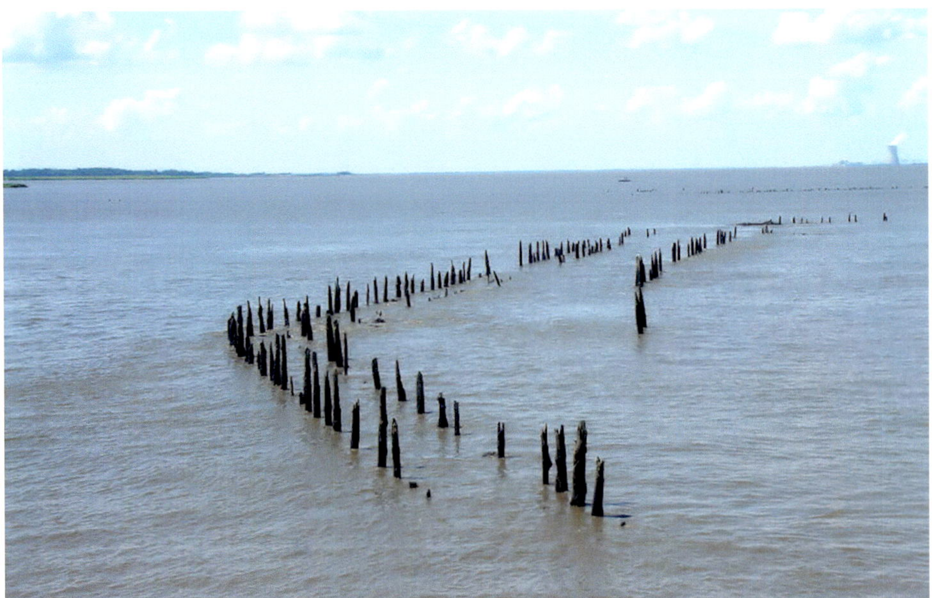

The city of Smyrna, Delaware, had a love-hate relationship with the railroad. When the Delaware Railroad was coming south, the city refused the train, saying they did not want the noise and dirt that came with the trains. So, the Delaware went west and created the town of Clayton, Delaware. This move left Smyrna isolated, so when a second chance came, the city took the opportunity. The line, however, proved to be little more than a novelty, and was abandoned to passenger traffic by 1928. While the freight line remained, it saw forty-one annual cars in its last decade, which ended in 1976. In 2019, a new water project revealed long buried rails along the line.

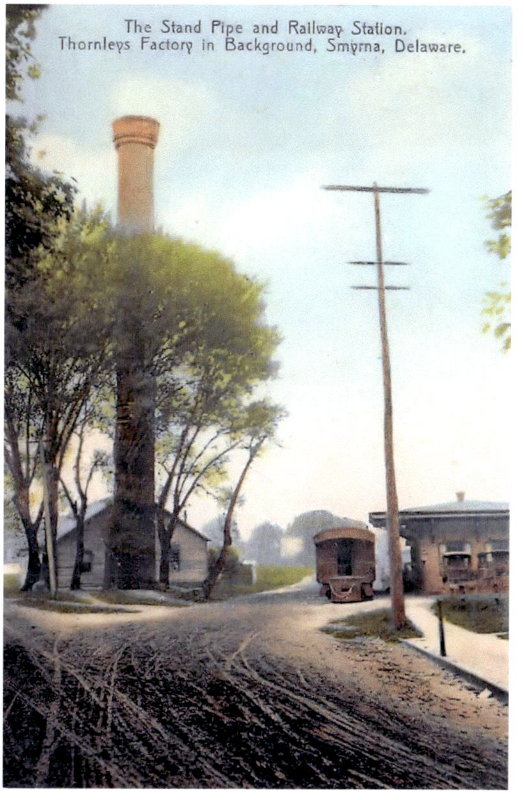

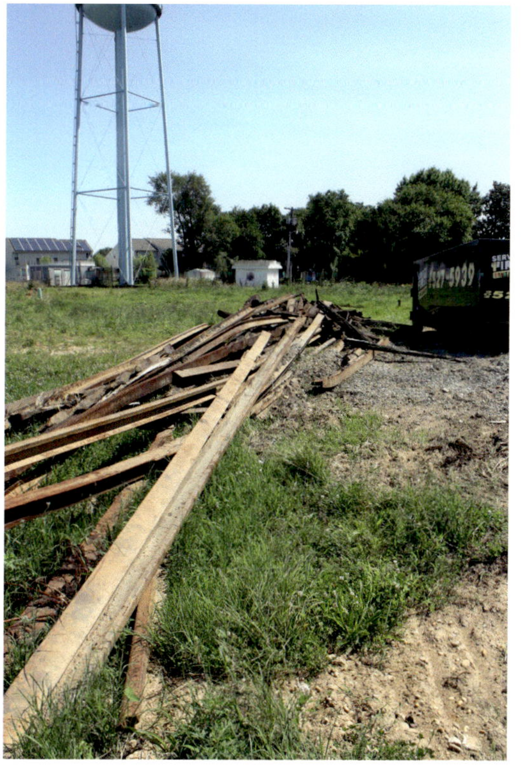

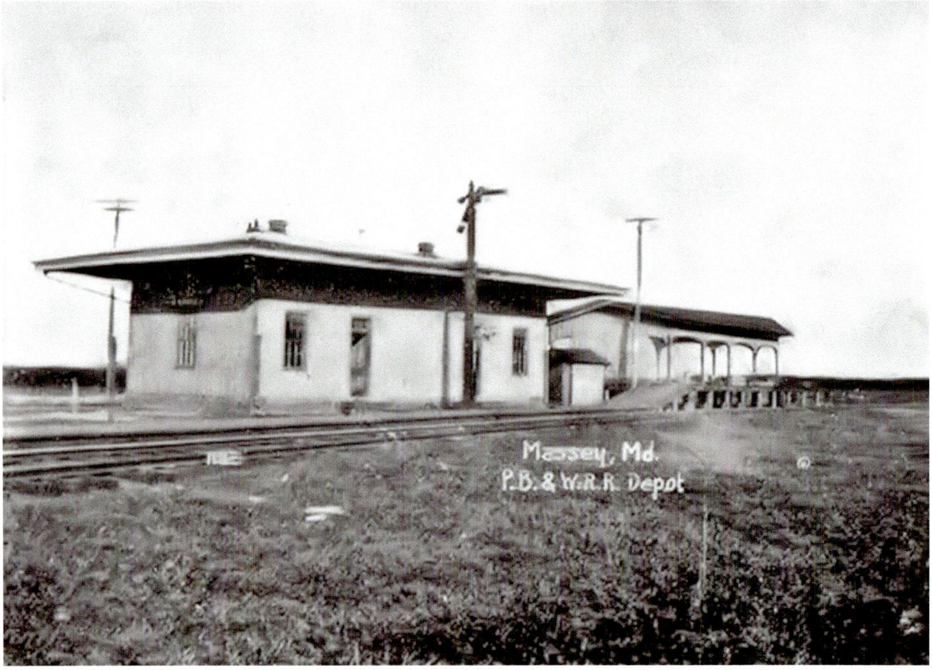

When the railroad came through, it created communities out of simple crossroads, as the book has mentioned. But occasionally, it created more. Massey, Maryland, is a great example of this. Not one, but two railroads converged here, creating a hub of activity. The Queen Anne & Kent Railroad was the first to come in 1856, with the Kent County Railroad arriving in 1868. Canneries, a large milk depot, and a sawmill all quickly sprang to life along the rail line. But the great dreams of this intersection never came to pass. The Queen Anne & Kent Railroad never made it to its goal of the Chester River and the Kent County Railroad did not reach its dreams of Rock Hall or Tolchester Beach. Massey still sees rail traffic from the Maryland and Delaware, a short line railroad company that handles several portions of former railroad lines all over Delaware and Maryland.

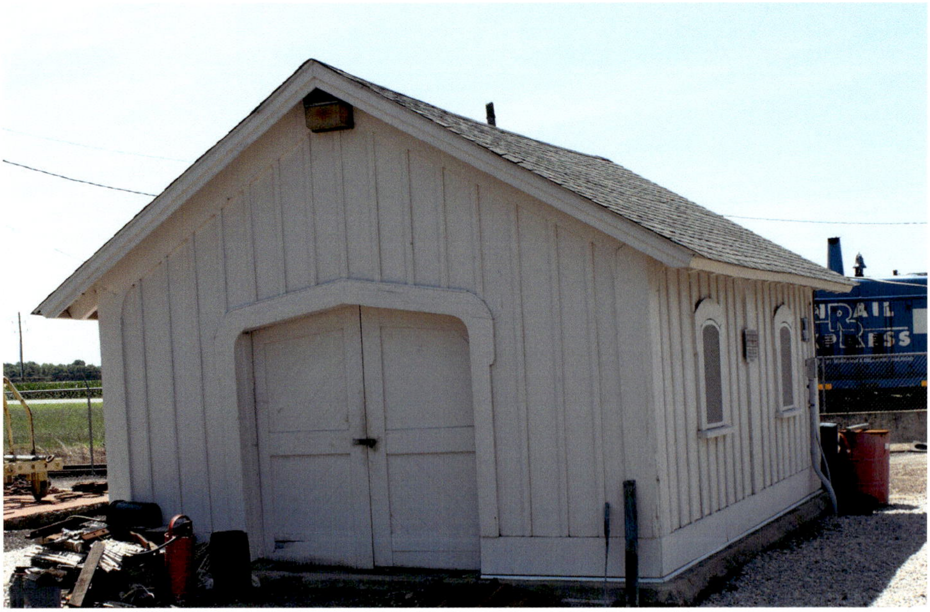

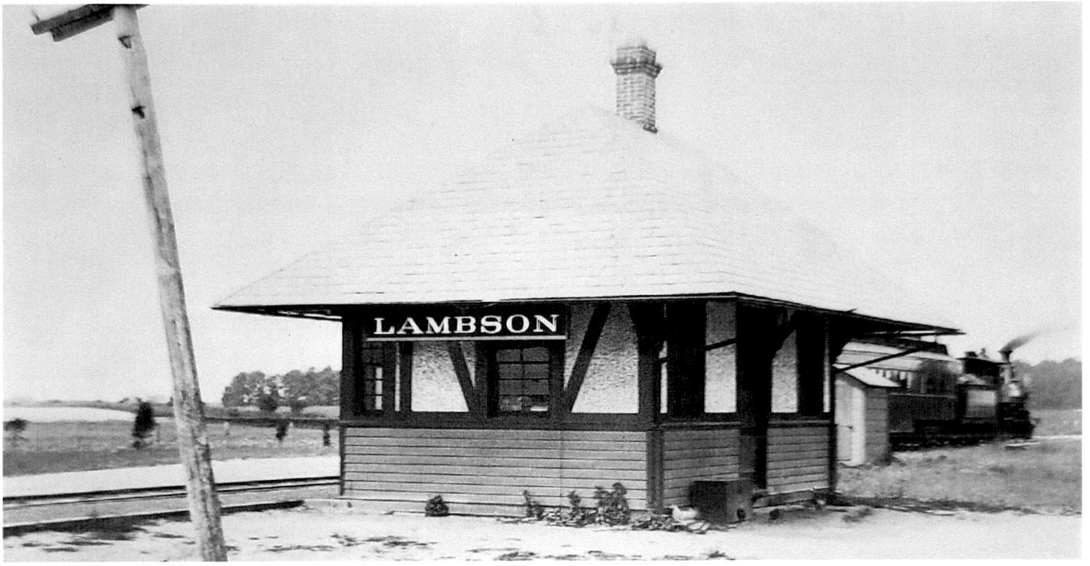

West of the Massey Junction, the Kent County Railroad continued on a due west course. While its goal was Chestertown and beyond, the rails passed through areas where cornfields and cows outnumbered citizens. Despite the small population, stations were constructed, more focused on freight than passenger travel. Lambson, Maryland, was the first of these stops leaving Massey. Even a small cannery was constructed in 1915, demonstrating the insatiable appetite of the country for the area's agricultural bounty. Still Pond, Maryland, was the next small hamlet to gain its train station. Not long after its construction, a cannery was constructed here as well. The area was the first in Maryland to allow women to vote in Maryland and today almost the entire area is under historic protection.

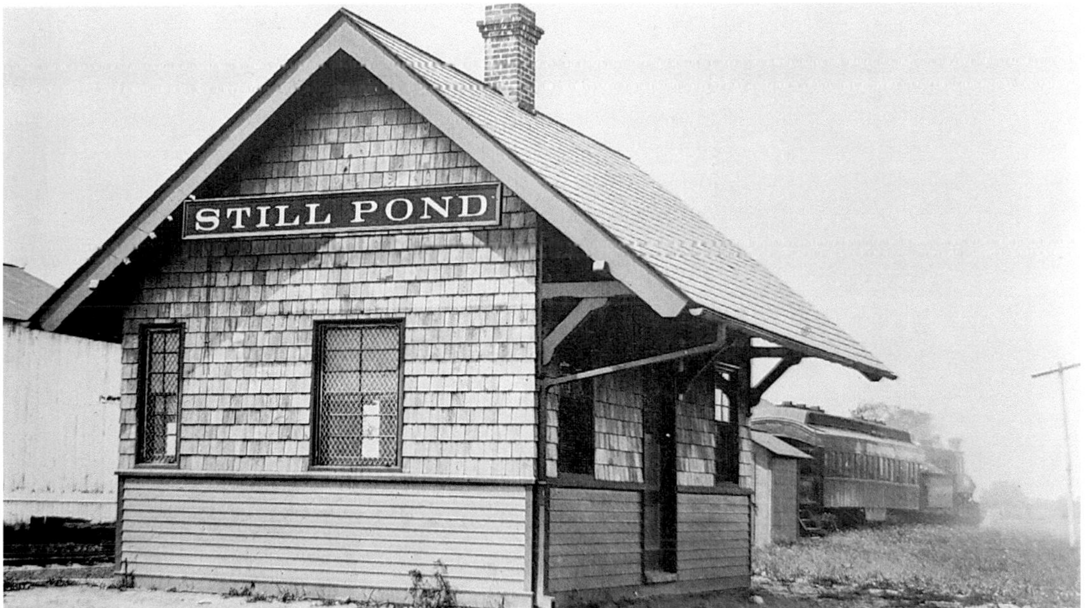

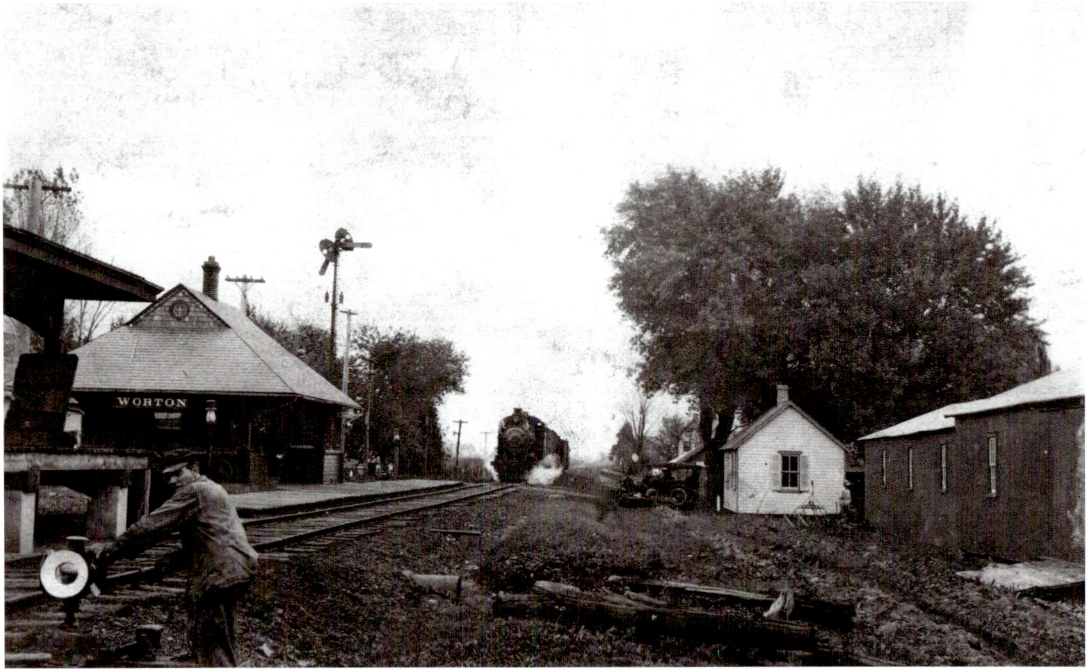

Worton, Maryland, was to be the junction that led the way to the Chesapeake Bay and its ships, passengers, and freight exchanges that would make the entire rail line a huge success. Leaving Worton, the line split, south going to the Chester River port of Chestertown, while its westerly trail would reach Rock Hall, Maryland. The line reached the goal of Chestertown, and its bigger dreams were never realized. The line to Chestertown was eventually removed, but Worton is the end of the line today for the Maryland & Delaware Railroad. *[Top photo courtesy of Jim Bowden & Keith Short]*

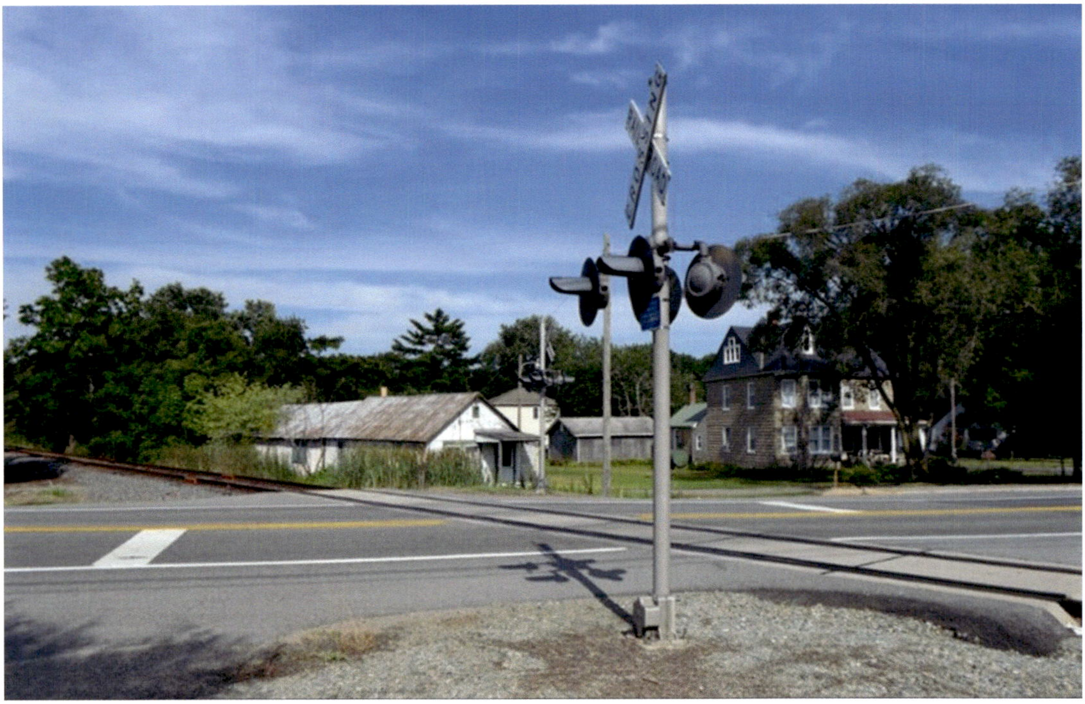

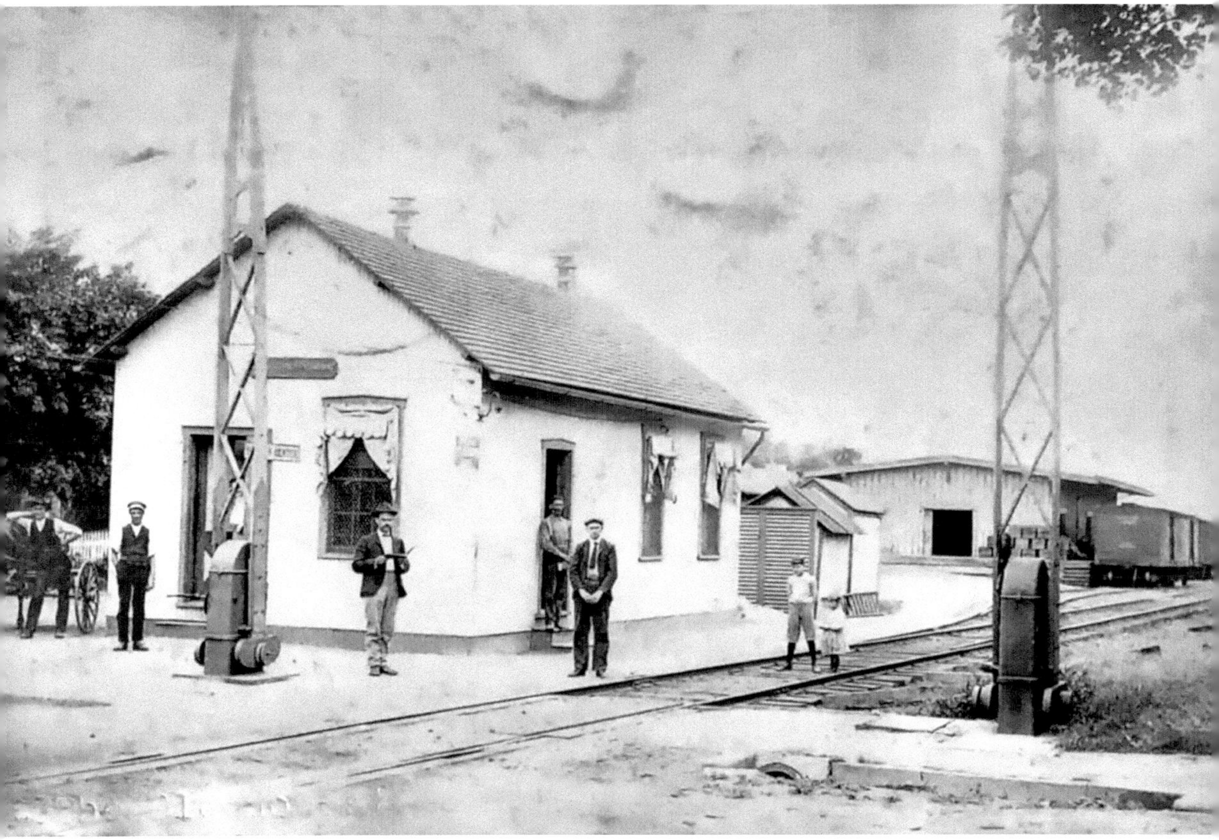

The city of Newark, Delaware, was home to three different railroads, each with its own station. While all three rails lines are still active in some fashion, one has mostly been abandoned. Originally known as the Pomeroy and Delaware City Railroad in 1872, by 1879 the company was bankrupt and reorganized as the Pennsylvania and Delaware. This group fared no better and the line was again sold, this time to the Pennsylvania Railroad. The changes were not yet finished as the PRR renamed the line to its most well-known name, the Pomeroy and Newark Railroad. The portion from Newark to Delaware City became a part of the Baltimore & Philadelphia Railroad, which was constructed by the B&O Railroad following its loss of the P.W. & B. to the PRR. Note the construction of the Baltimore and Philadelphia was so costly, it caused the B&O Railroad to file for bankruptcy in 1896.

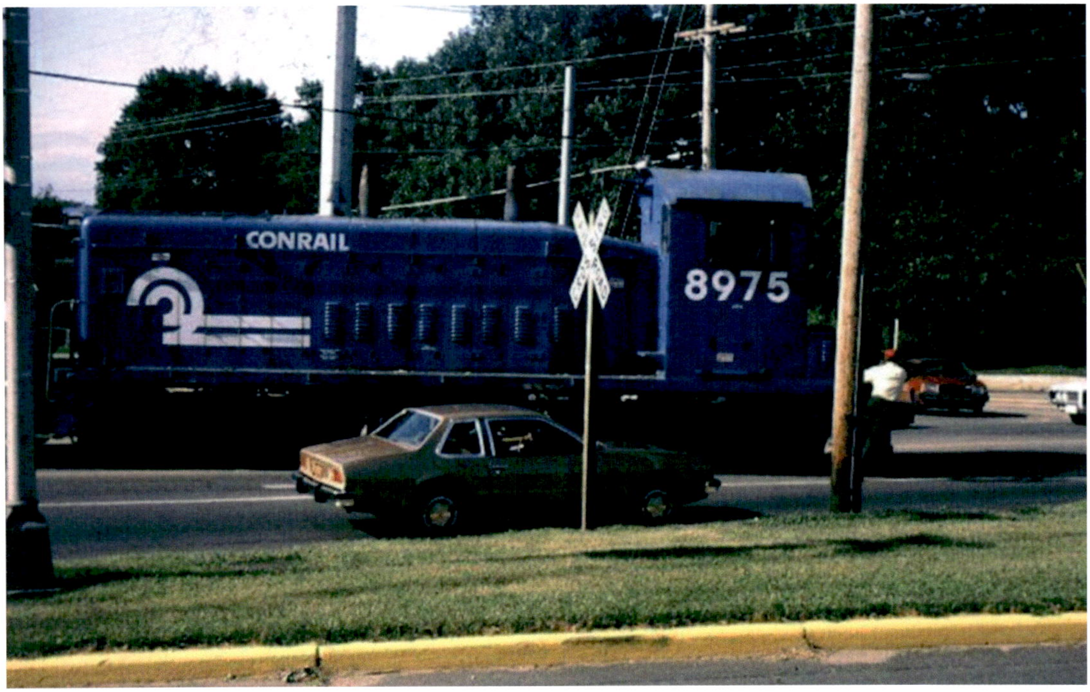

The Newark portion of the line stayed in service well into the Conrail days. While no through traffic traveled the rails, several businesses such as Curtis Paper, Newark Lumber, and Dayett Mills used the railroad for local delivery to and from their locations. All traffic stopped using the track in 1984 and it sat abandoned for over two decades. Recently, the area has been turned into a bike trail, but the history of the area was not forgotten as this old switch serves as a reminder of the now vanished rail line.

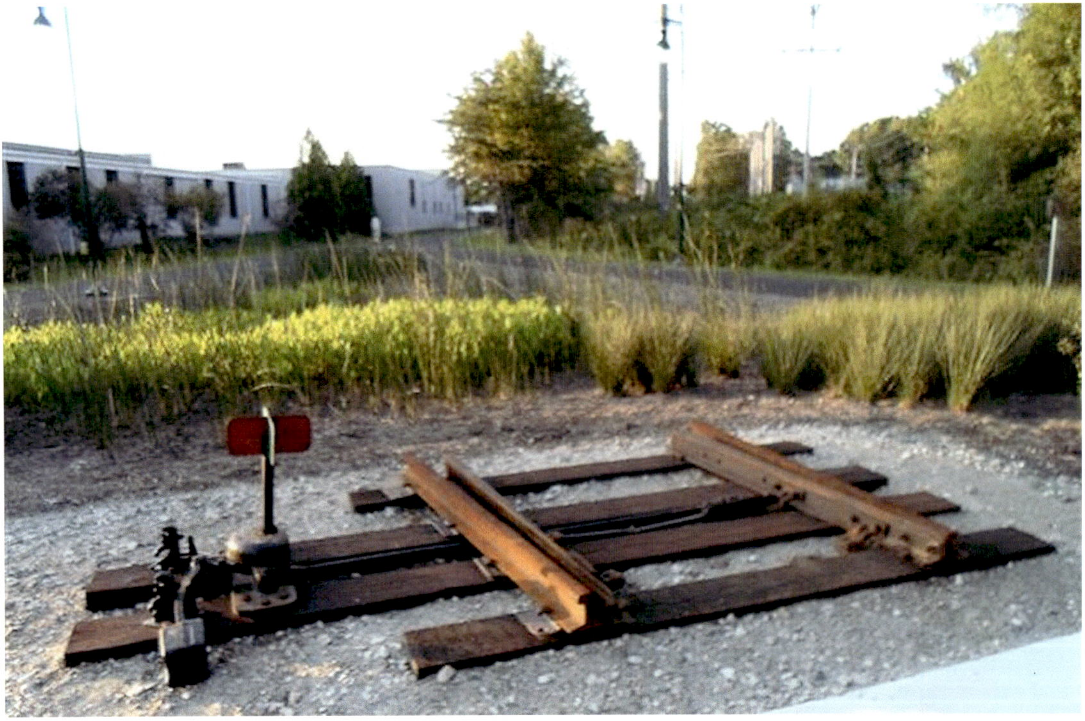

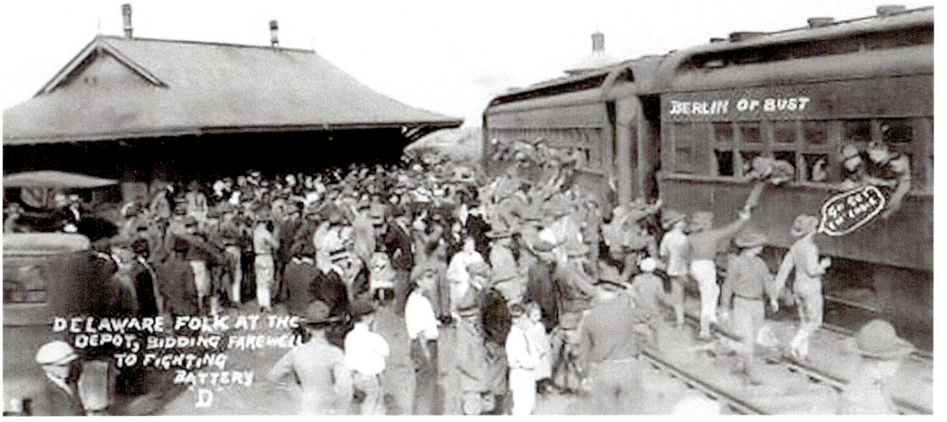

The southern terminus of the Newark and Delaware City Railroad was Fort DuPont, which lies just to the south of Delaware City, Delaware. The fort was a key instillation during WWII. During the war, frequent passenger trains would come and go from the station, carrying military men to major cities where they boarded ships for their travels across the Atlantic. [*Photo courtesy of Jim Bowden & Keith Short*]

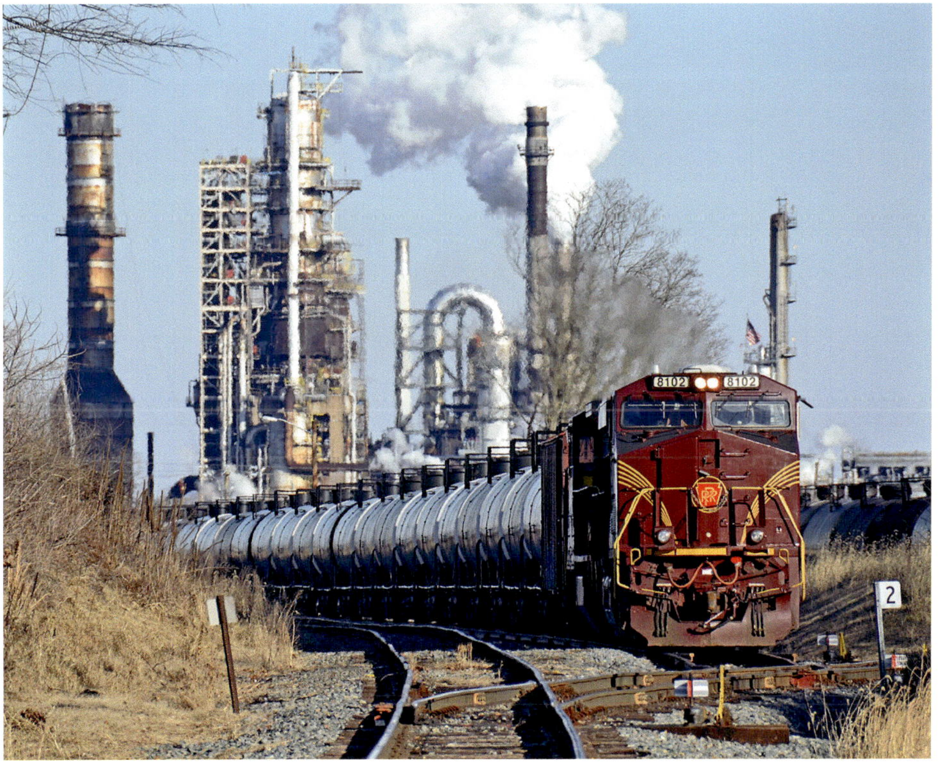

While most of this old short line was destined to be a failure, a portion remains today and is more active than ever before. Norfolk Southern travels these rails to and from oil refineries that are situated just north of Delaware City. Train enthusiast today can see well over 100 tanker cars being moved to and from these facilities daily.

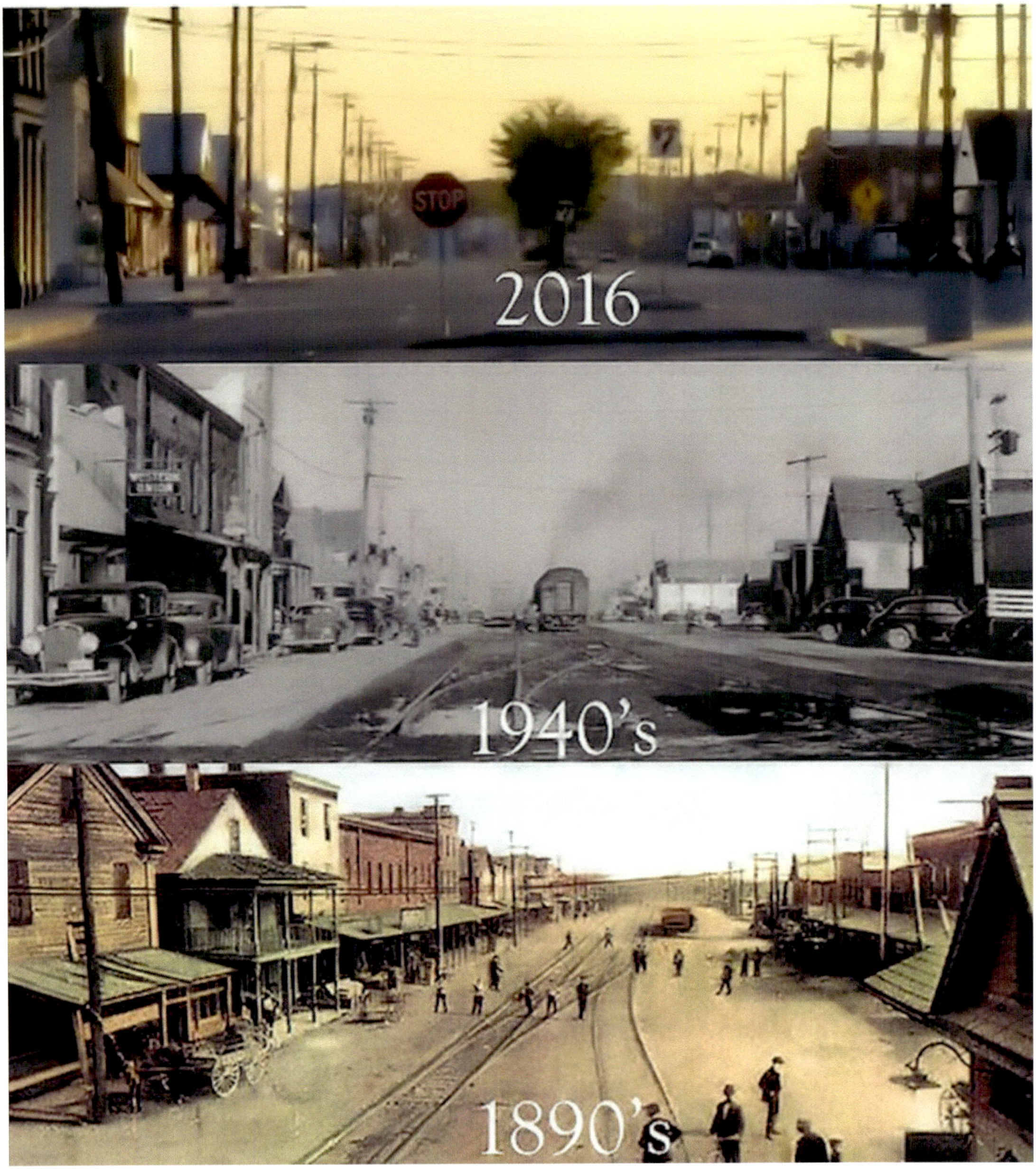

Produce was not the only major commodity that built the railroads on Delmarva. Further south into Maryland, seafood was king. For many years, Crisfield was the leading oyster producing city in the entire United States. The combination of the railroad and the oyster helped fuel the development of secondary industries in the city. By 1904, the City of Crisfield was the second largest city in the State of Maryland (the first being Baltimore) and one of the finest seaports in the country. It quickly became known as the "Seafood Capital of the World."

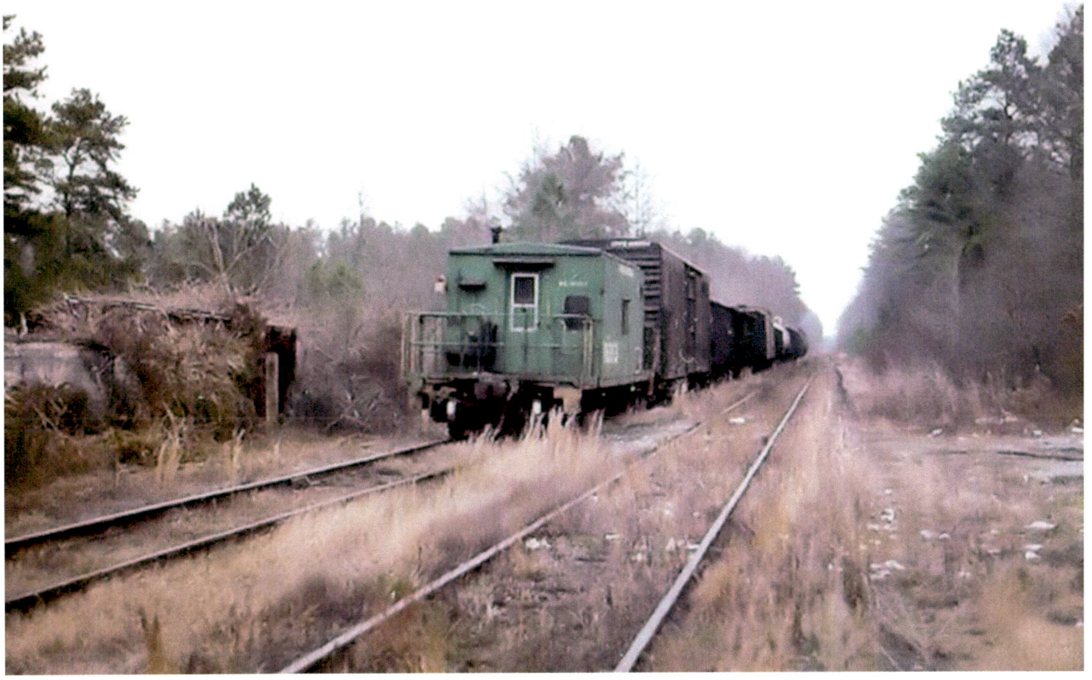

In 1884, another group of railroad pioneers saw the bounty of the lower Eastern Shore of Virginia and beyond. William Scott and Andrew Cassatt saw the great opportunity and seized upon the chance. Little was in the area as most villages were along the coast. So Cassatt ran the line right down the middle of the area. Towns such as Lecato, Virginia, were created overnight as stations were built along the new railway. Each town became a hub of activity as locals shipped their goods via the new railroad. [*Top photo courtesy of Ed Sharpe*]

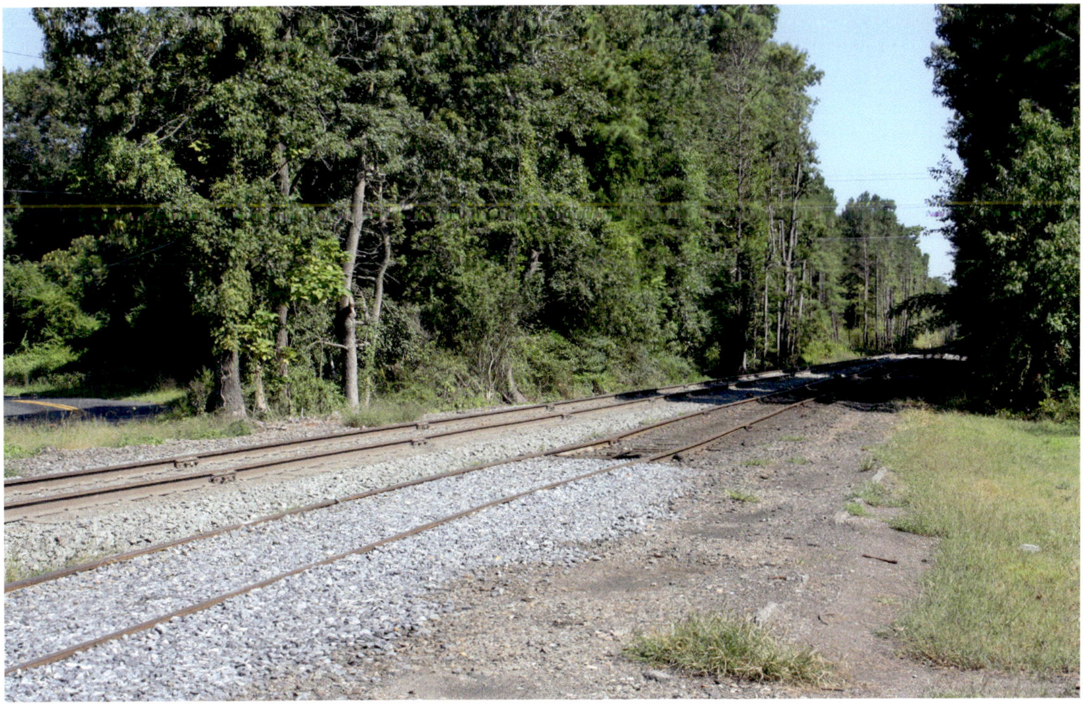

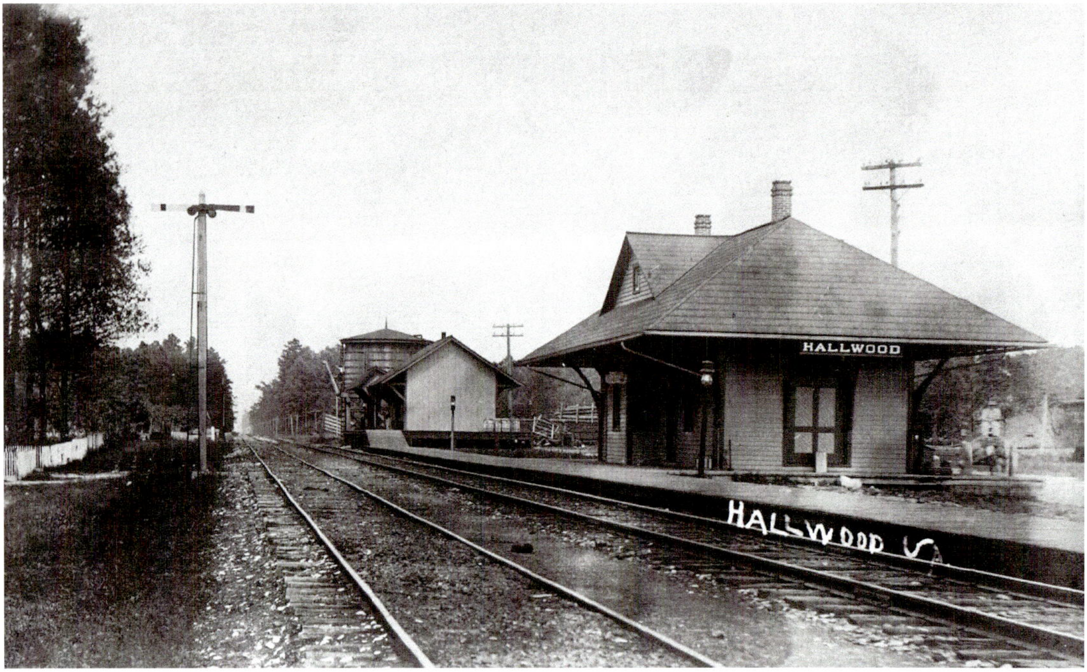

Over a dozen new towns sprang up as the New York, Philadelphia, and Norfolk Railroad made its way down the spine of the eastern shore of Virginia. Prior to the railroad coming, most people lived in seaside communities of less than 100 people. The valuable and fertile farmland quickly changed all of that. The Eastern Shore of Virginia Produce Exchange was created on September 30, 1899. By 1910, the exchange saw average sales for its farmers in the area of between five and ten million dollars a year. Property values went from ten dollars an acre to well over 100 dollars in that same time. [*Photo courtesy of Jim Bowden & Keith Short*]

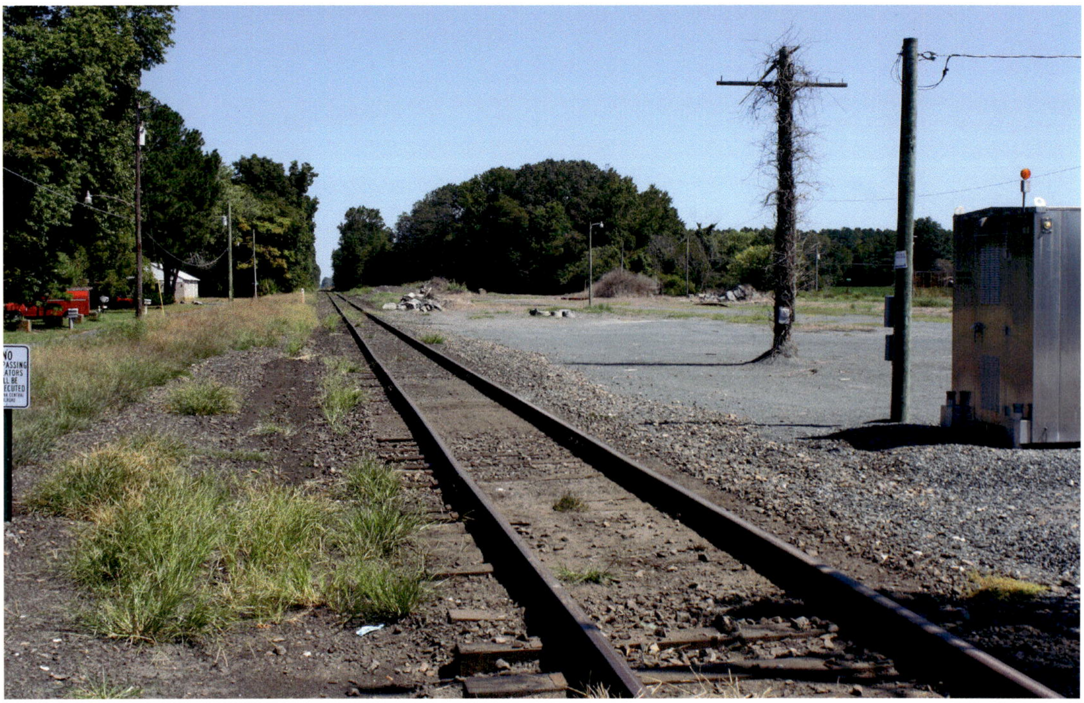

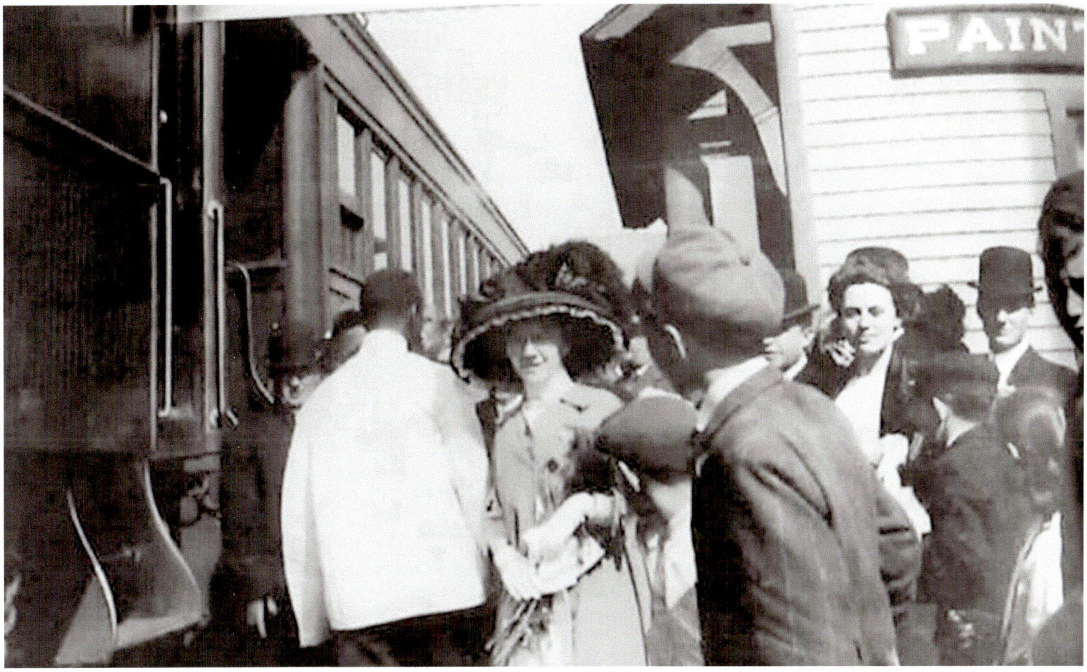

Painter, Virginia, was established in 1784 and at the time was called Mappsburg, Virginia. The small town was located closer to the coastline at the time and remained a small seafood port until 1908. While the town of Painter had been established with the railroad, the two areas combined and grew Painters population to over 1,000 residents. The crowd can be seen gathering to board a passenger train. This was a common sight until after World War II. Today, the old rail is covered with weeds and the platform curbing that was once crowded with those passengers sits silent. [*Top photo courtesy of Jim Bowden & Keith Short*]

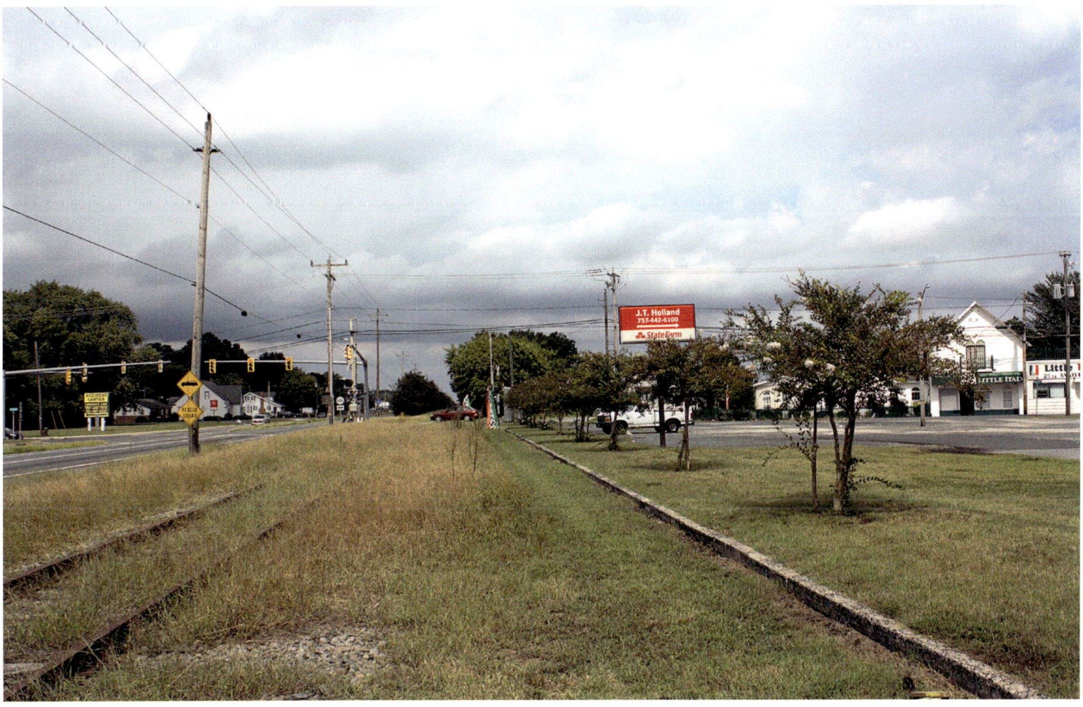

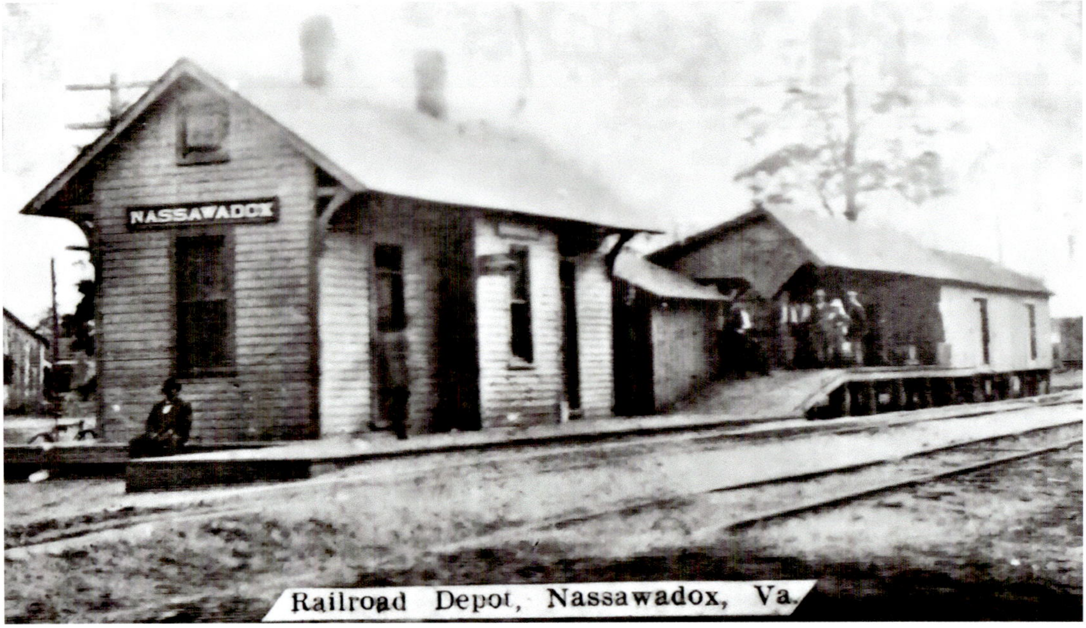

A quote from a visitor to the area in the early 1900s: "They are engaged in feeding populations numbering several millions, from 200 to 500 miles northward. The rapid trains for the quick delivery of produce go as far as Boston, and in some cases to Canada. In twelve hours the fresh and tempting fruits and vegetables are delivered in New York, in twenty hours in Boston, and in thirty hours in Montreal." Nassawadox, Virginia, was representative of this quote. It had bountiful farmland and water teaming with seafood to feed millions. Today, just a restored caboose sitting on rails reminds the passerby of the once tons of freight and thousands of passengers that the rails once served. [*Top photo courtesy of Jim Bowden & Keith Short*]

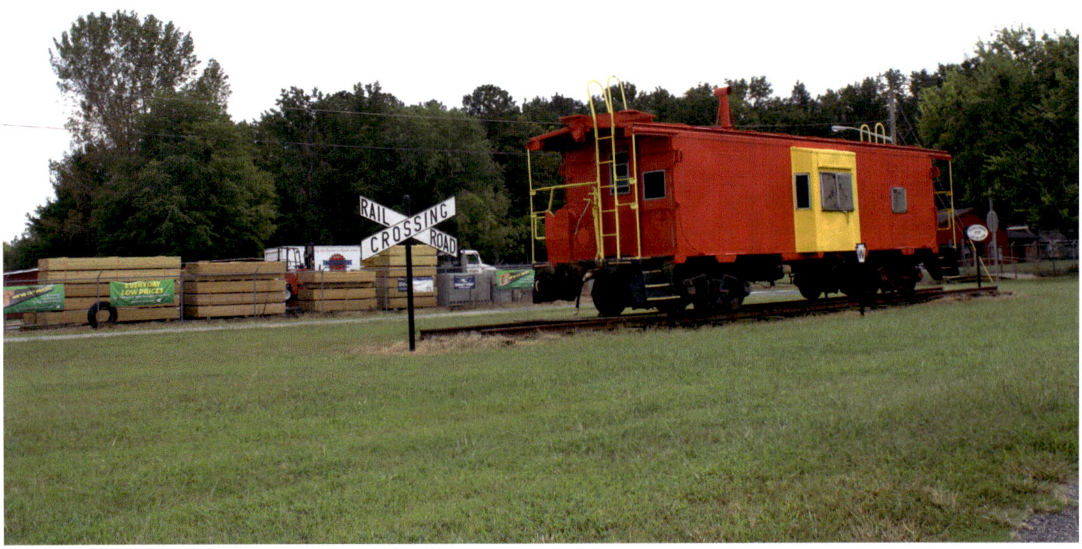

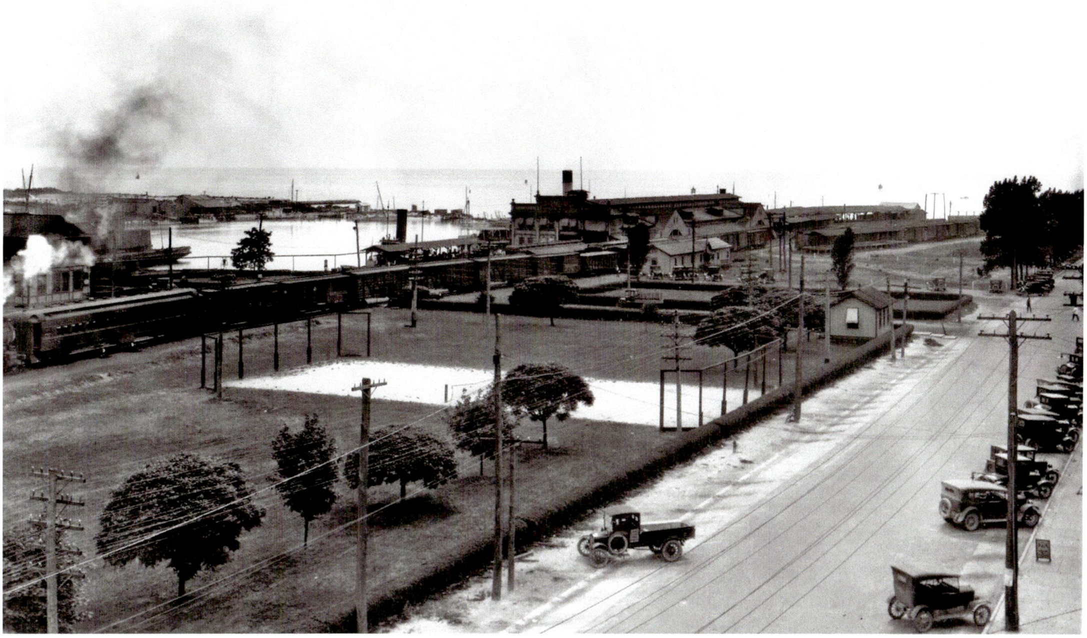

The final stop on this line to the south was Cape Charles, Virginia. Literally constructed from a marsh that could barely float a toy boat, the entire area was turned into a bustling railroad city. Passenger service was separated from the freight terminal and for a short period, you could even continue to sleep in your berth as your entire railcar was loaded onto a barge for transport across to Norfolk, Virginia. An average of almost 200,000 passengers a year used the ferry service to and from Norfolk. The last passenger train in the area was in 1958. The area today has been cleared and is awaiting development. [*Top photo courtesy of the Eastern Shore Railroad Museum*]

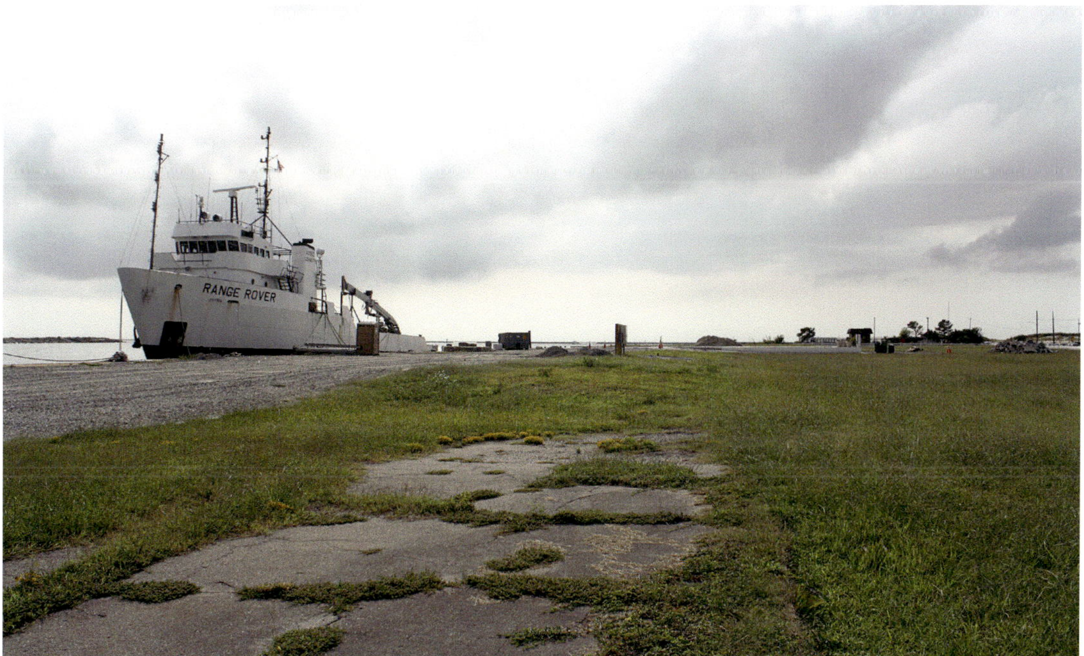

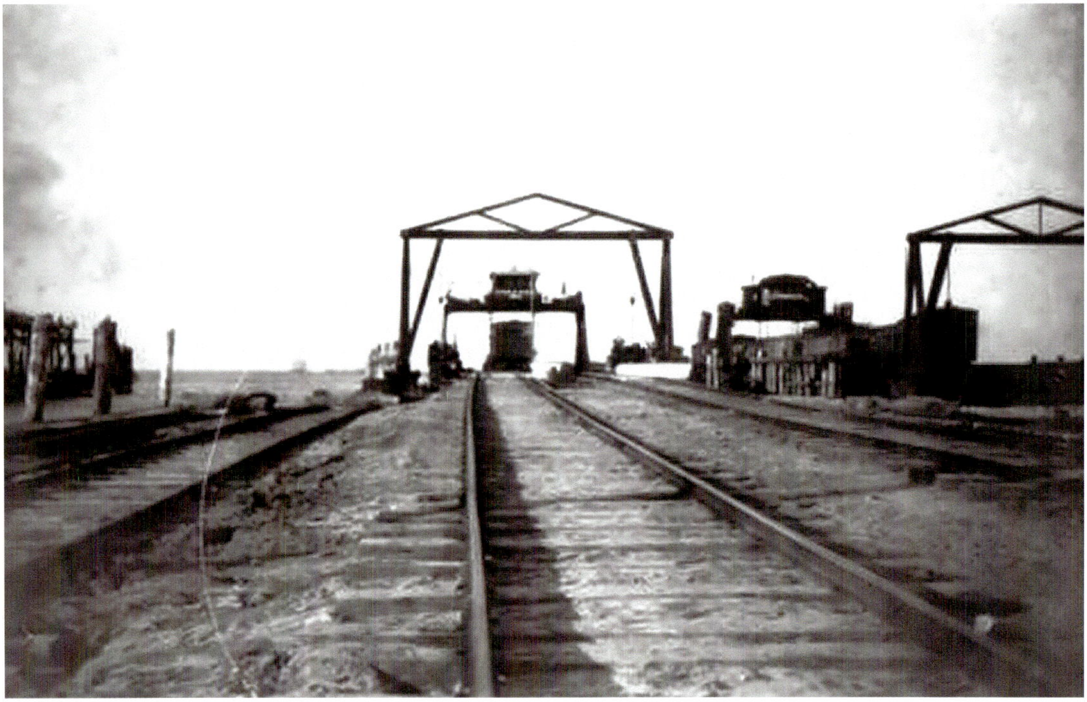

While passenger traffic was handled mostly by the streamer ferries, freight was the main reason the railway was created. By the early 1900s, the NYPN saw approximately three million short tons of freight with a value of over 116 million dollars. The barges were an engineering feat designed by Cassatt. Prior to his design, no barge could carry as much weight as those he designed for the railroad. While passenger services shrank, the freight service continued well into the 1960s at which time the Chesapeake Bay Bridge Tunnel was built, slowly bringing to an end the once massive operation. [*Top photo courtesy of Hagley Museum*]

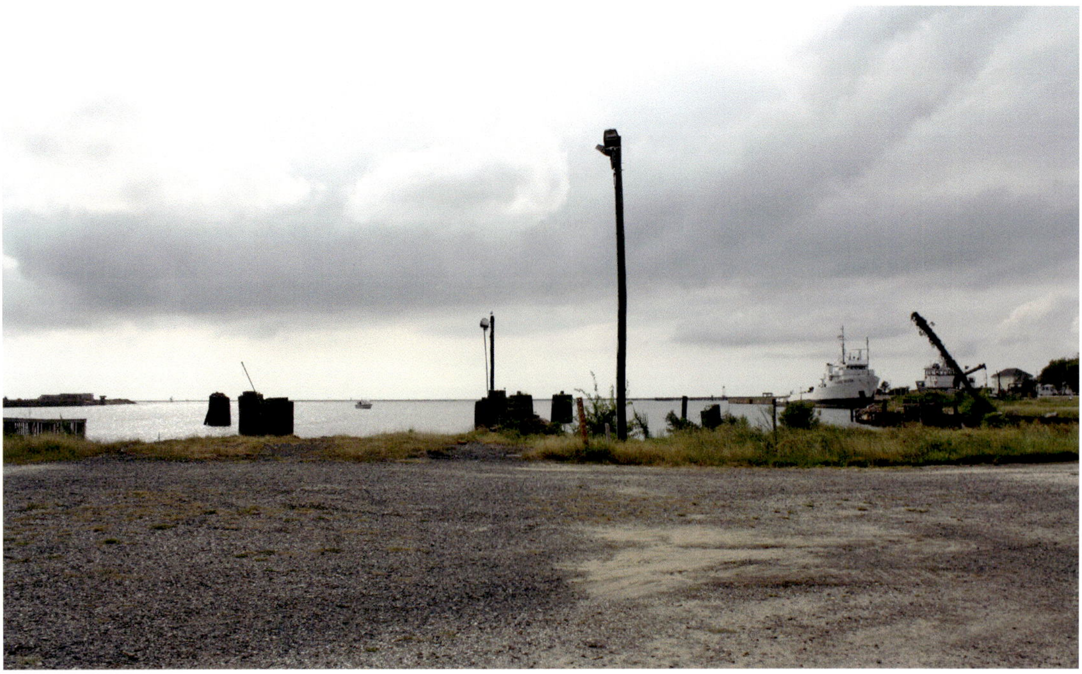

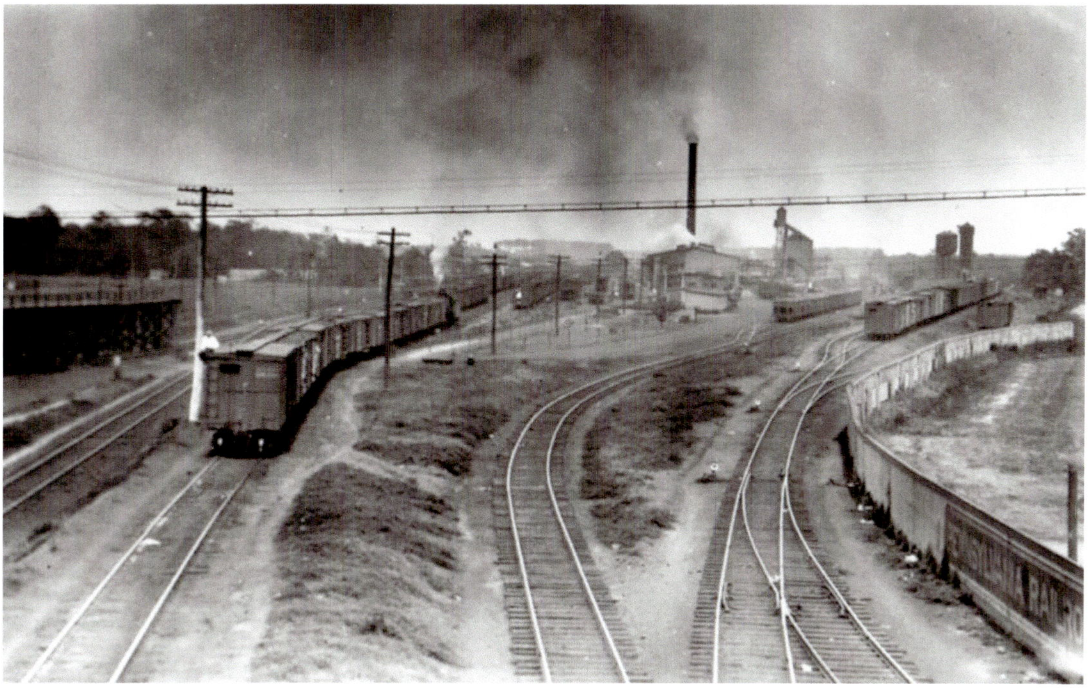

The rail yards at Cape Charles were vast in both size and movement. The twenty-four-hour-a-day operation would see on average of over 1,000 freight cars a day moved around the yard. Maintenance was non-stop on all manner of railroad equipment. The area grew so active, the railroad constructed its own power plant to provide electricity to the area. Offices, bunk houses, roundhouses, and repair shops were built to handle the ever-growing hustle and bustle of the NYPN. As with so much else, as the traffic dwindled, the area fell into disrepair. In 2018, the entire railyard was removed and sold for scrap. [*Top photo courtesy of Ed Sharpe*]

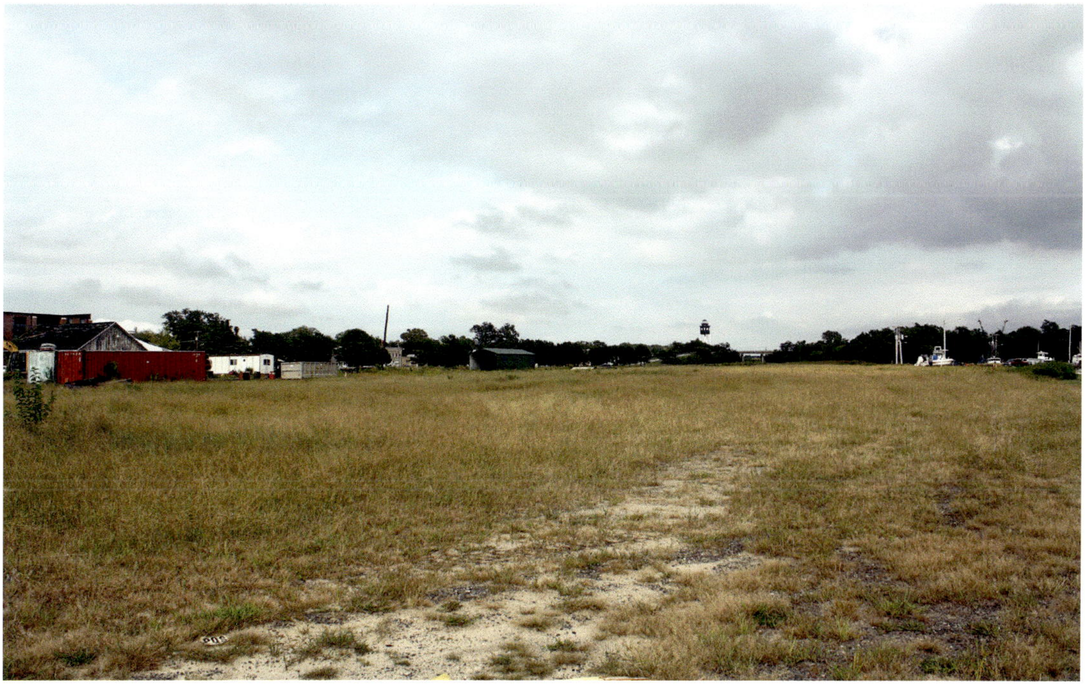

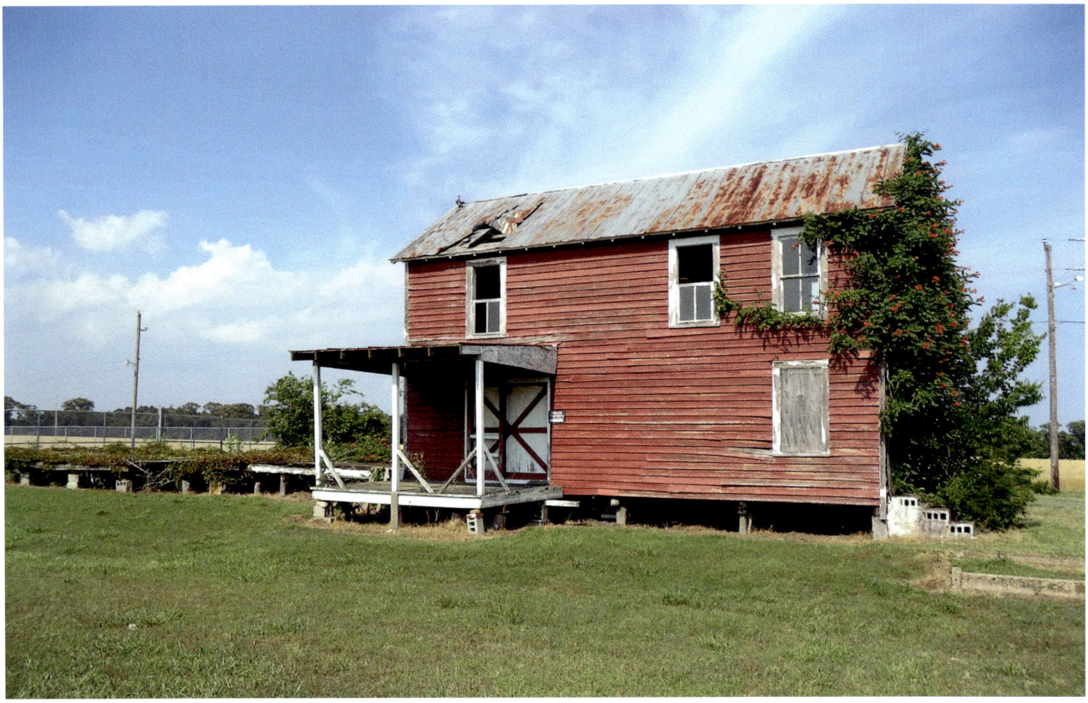

The Cape Charles Railroad was incorporated in the state of Virginia on March 24, 1906. It was a subsidiary of the New York, Philadelphia & Norfolk Railroad (N.Y.P.&N.) and was set up that way to avoid legal complications. The purpose of the Cape Charles Railroad was to build a spur of about 12 miles down to the cape to Kiptopeke Virginia. In 1915, the area that the Cape Charles Railroad served produced 520,000 barrels of potatoes.

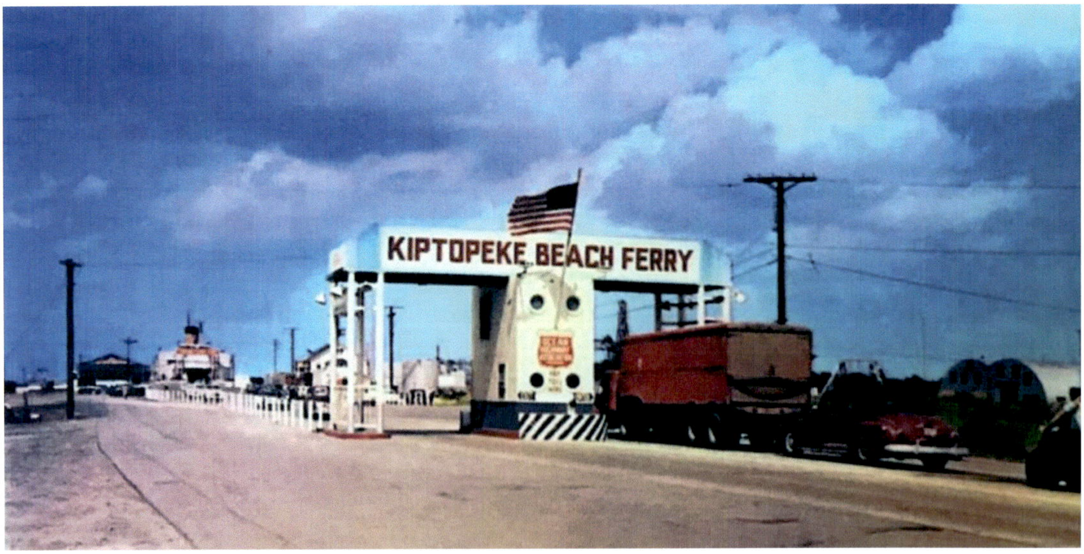

The Little Creek-Cape Charles Ferry was a passenger ferry service operating across the mouth of the Chesapeake Bay from the 1930s until 1964. Beginning in the 1940s, the ferry began accommodating vehicles as well as passengers. During its peak, the Little Creek Ferry operated ninety one-way trips each day with seven vessels. In 1949, the northern terminus for ferry service was moved from Cape Charles to Kiptopeke, where a new pier was completed in 1951. The area was created using nine concrete surplus ships from WWII.

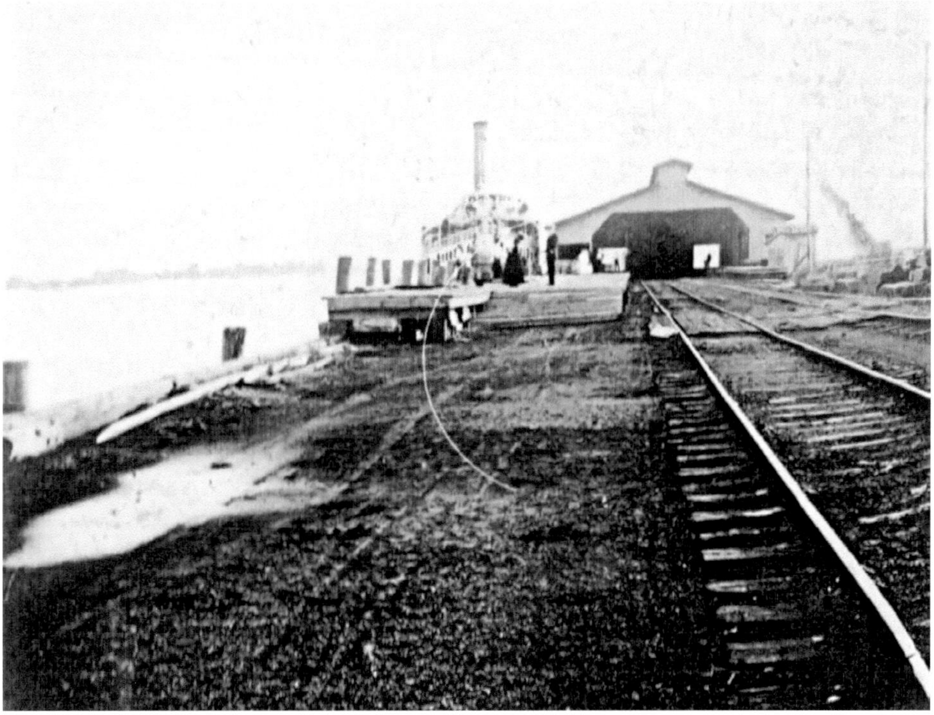

While a railroad now ran north and south across the Delmarva Peninsula, the cities of Baltimore and the surrounding areas wanted access to the area with a train that ran east and west. So, in 1886, the Baltimore and Eastern Shore Railroad was created. Its first challenge was to build a landing and pier that could handle both passenger and freight traffic. Clairborne, Maryland, was already a ferry terminal, so the railroad purchased the rights to the site. [*Photo courtesy of Jim Bowden & Keith Short*]

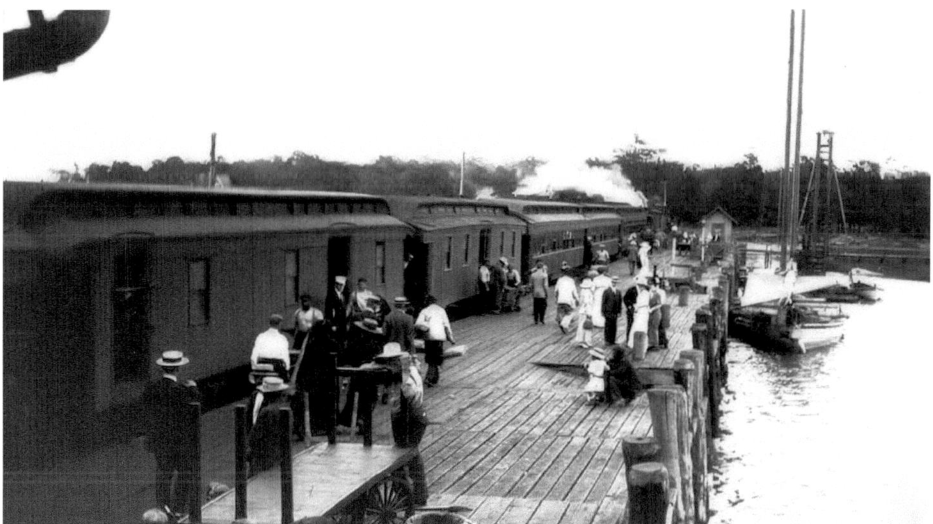

The railroad sadly saw no financial success and was sold into bankruptcy just four years after it began. The railroad that followed became one the most beloved ever to ride the rails on the Delmarva Peninsula. The Baltimore, Chesapeake & Atlantic Railroad quickly turned the line profitable and by 1912 was running several ferries and trains daily.

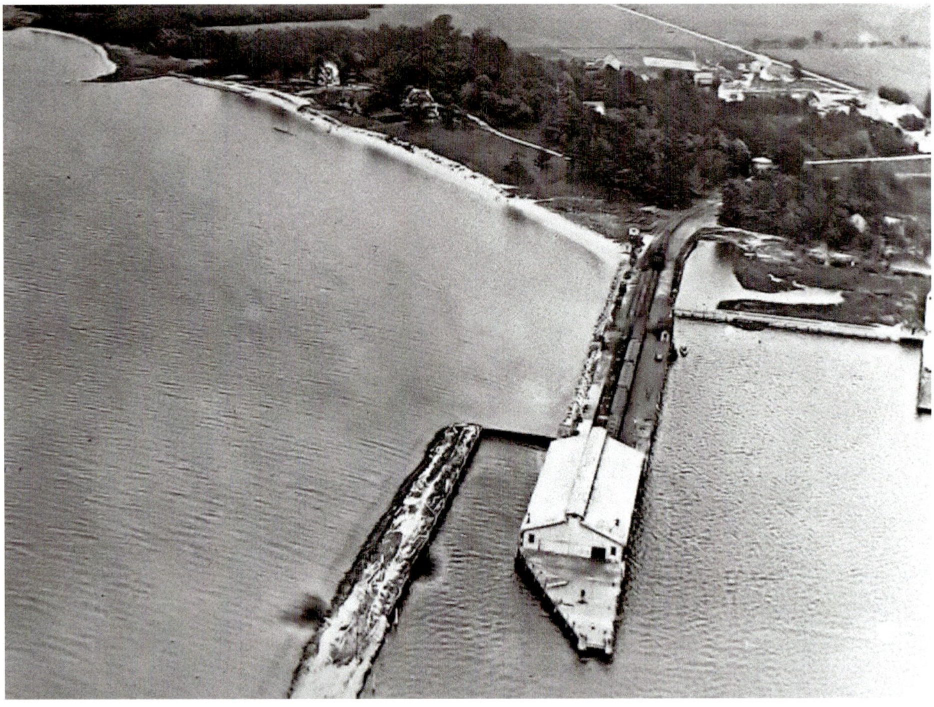

Eventually, even the "Black, Cinder and Ashes" as the line had become affectionately known by, would succumb to having to support its sister line, the Maryland, Delaware and Virginia, and was sold in 1928. The Baltimore & Eastern Railroad, with backing from the Pennsylvania Railroad, would assume control of the line. The pier at Clairborne continued to move freight for two more decades.

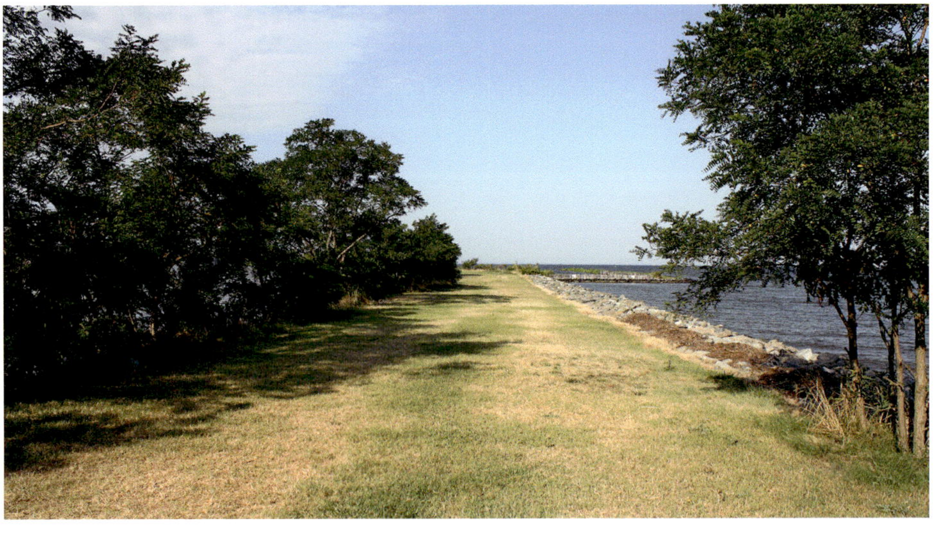

The strip of land the train used as a rail bed, and the pier was constructed upon, sits today as a small park. A public fishing dock and pier exist as well. Homes now look out upon the lazy river as the current rolls by. Seagulls now perch on the pylons that once served as the bulkheads for the pier. If you walk down toward the water's edge, you can see an old railroad tie peeking out from the sod. Closing your eyes, you can hear the steamboats arriving, the trains leaving the station, and all the hustle and bustle of the early 1900s.

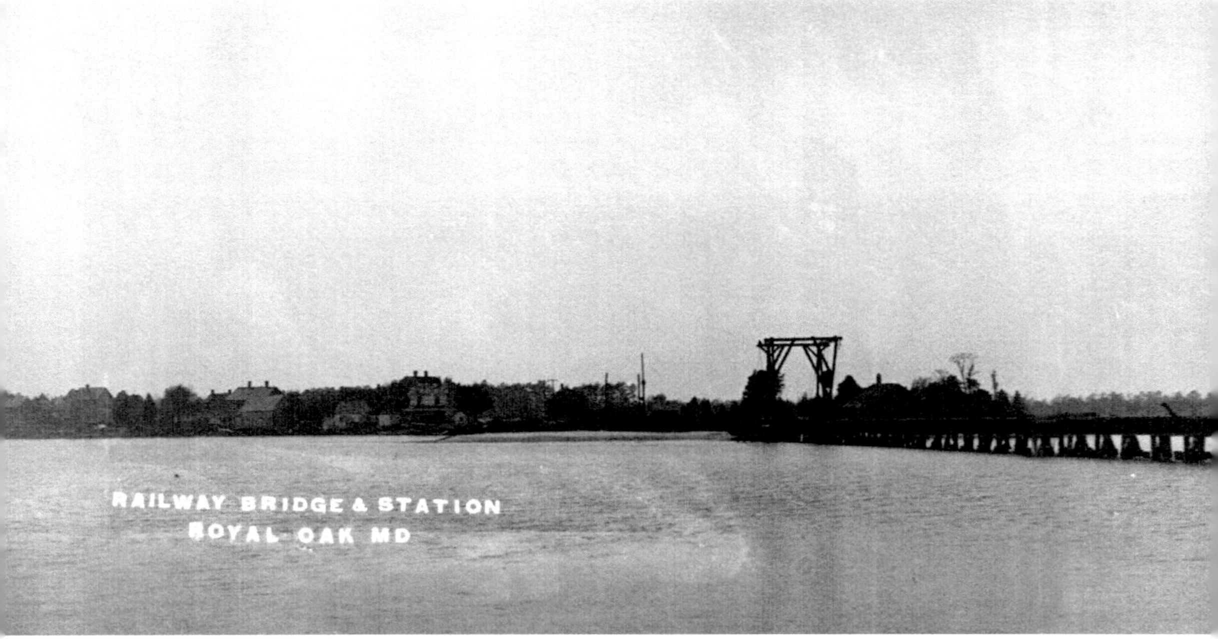

Above: While the bay crossing for the BCA was accomplished by the steamships, it was by no means the only major body of water that stood in the way of the train traveling to the east. The Miles River at Royal Oak, Maryland, was another waterway that posed a challenge for the BCA. A drawbridge was chosen as the method to span the opening from the river to Oak Creek.

Right: Rail traffic on the line continued until the early 1970s when it was finally abandoned. The drawbridge remained in place and the rest of the rails were removed. Boards were put down and the structure became a fishing pier for locals.

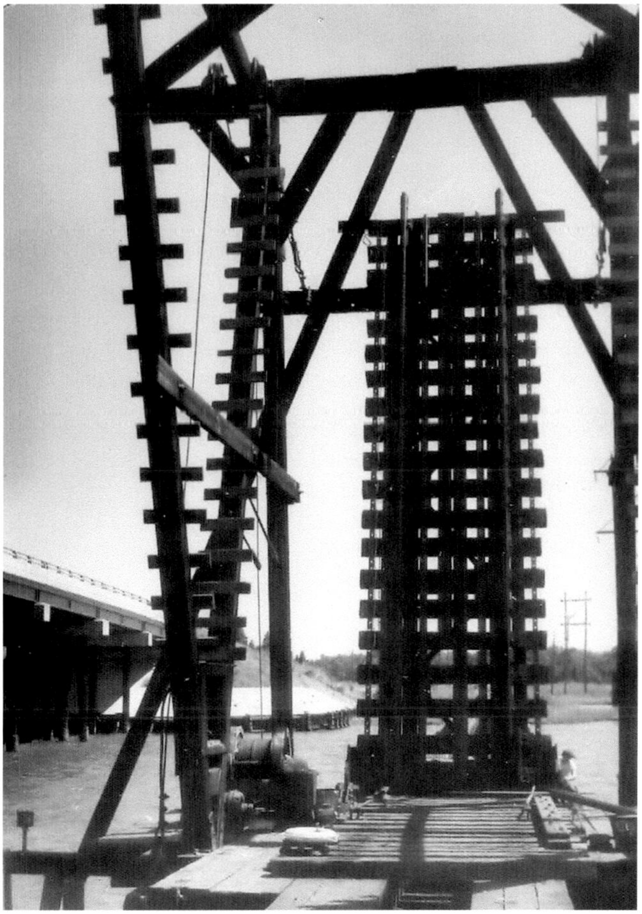

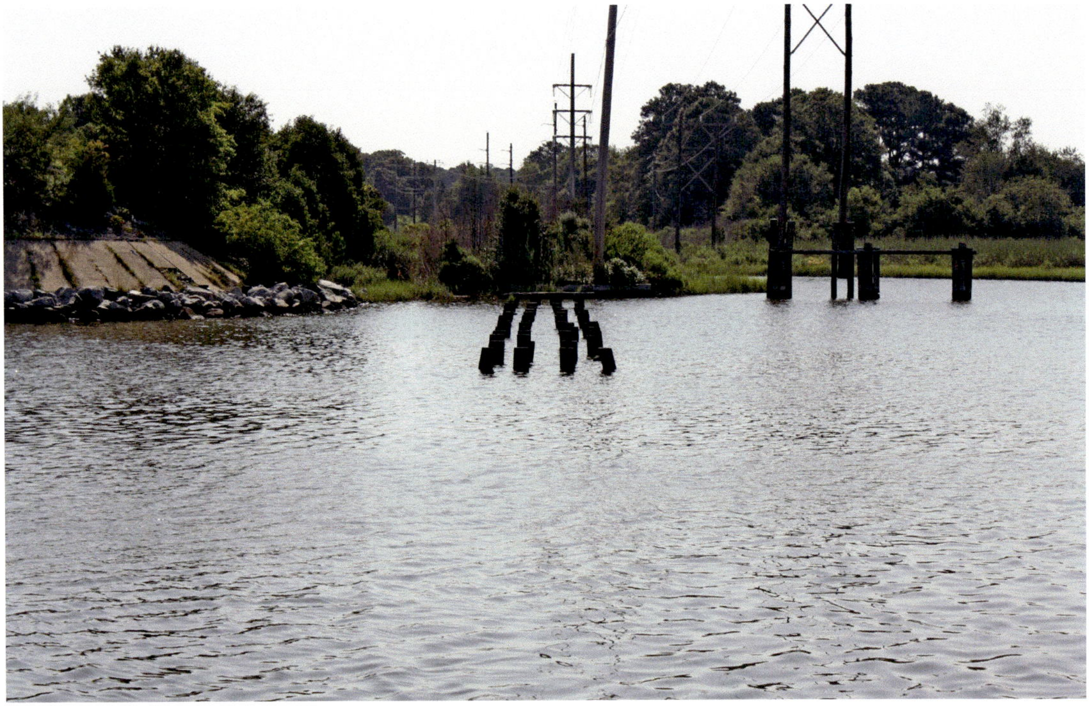

Today the rails and bridge have been removed and a new fishing pier has been installed on the west bank of Miles Creek. The piers for the old drawbridge can still be seen against the east bank, but slowly the water is reclaiming the last remnants of the structure.

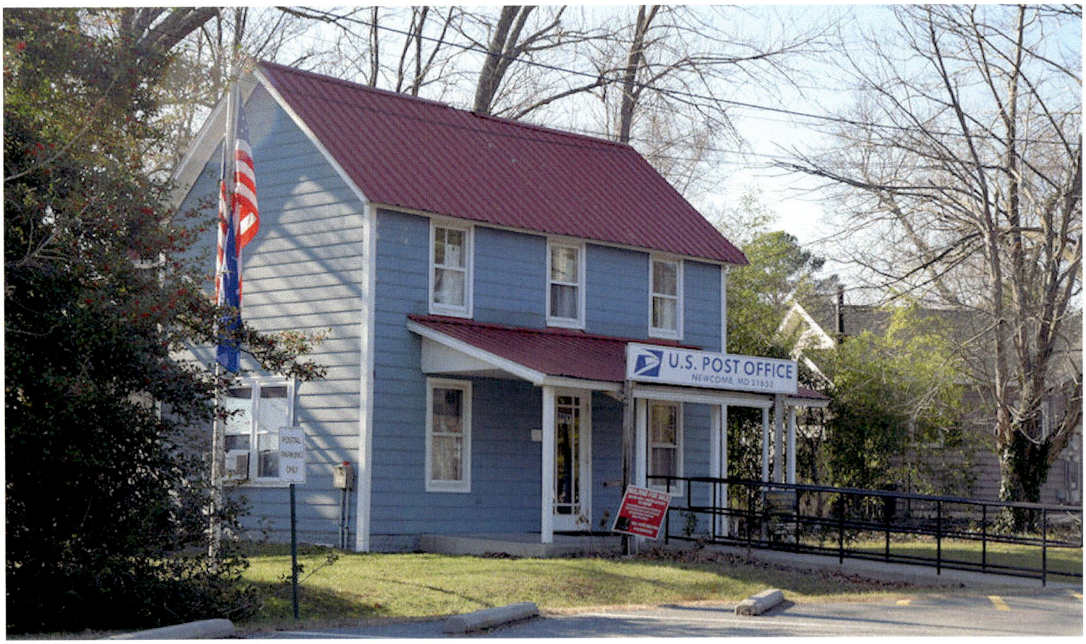

The actual nearest station to the Royal Oak Bridge was Newcomb, Maryland. The station was typical of the period. A two-story station would be built in some areas, with train work on the first floor and the station master and his family on the second floor. The station was relocated and today serves as the post office for Newcomb.

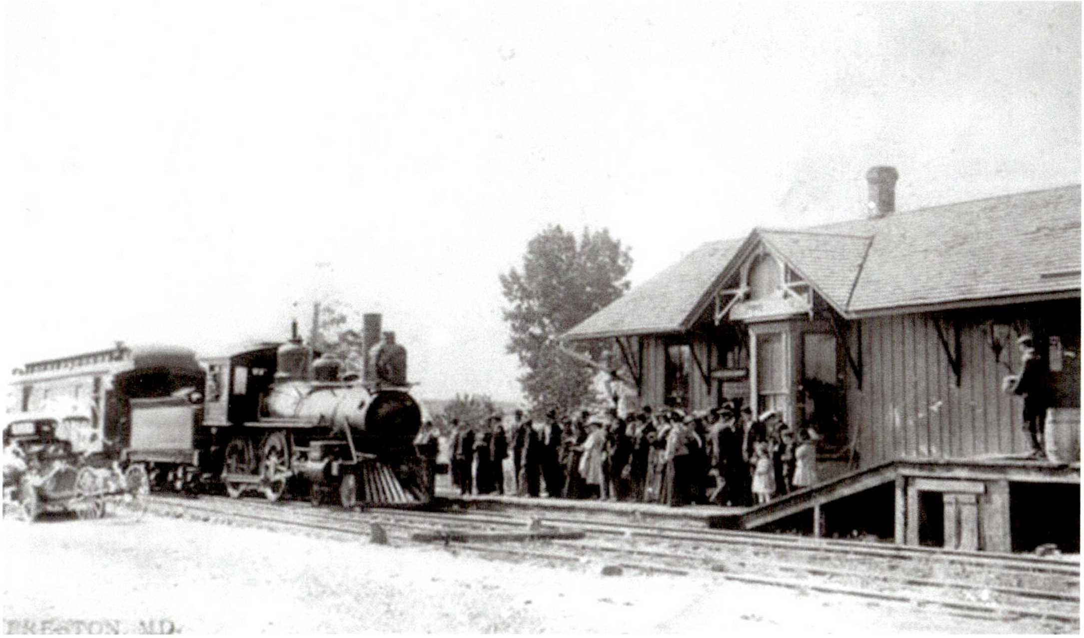

Preston, Maryland, was typical of so many small towns on the Delmarva Peninsula. Starting as a location for a Methodist church, the railroad boosted the local economy and subsequent growth. Four daily passenger trains and two freights stopped here. During the years before and after World War I, it was common for twenty to twenty-five freight cars a day of canned tomatoes, apples, wheat, and watermelons to be shipped during harvest time. A 1908 edition of the town newspaper reported that one-tenth of the total output of tomatoes was packed in Caroline County and that Maryland packed five-twelfths of all that was canned in the whole country. At one time, sixteen canneries existed within seven miles of Preston, and three were located within the town near the railroad. Some of these warehouses remain today, although the canneries are now gone, as is the railroad. [*Top photo courtesy of Jim Bowden & Keith Short*]

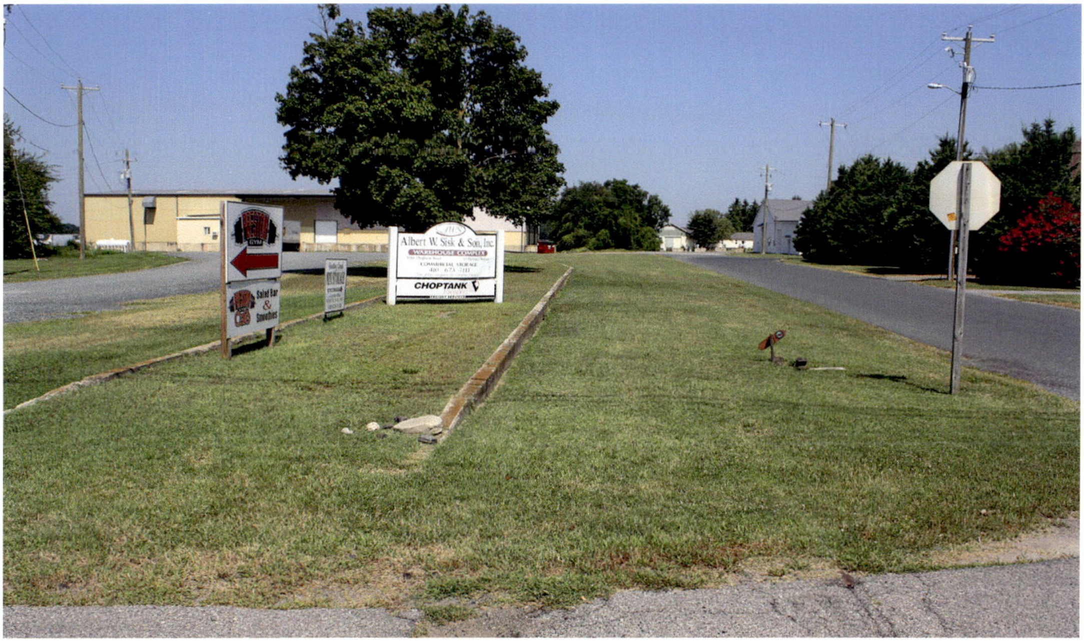

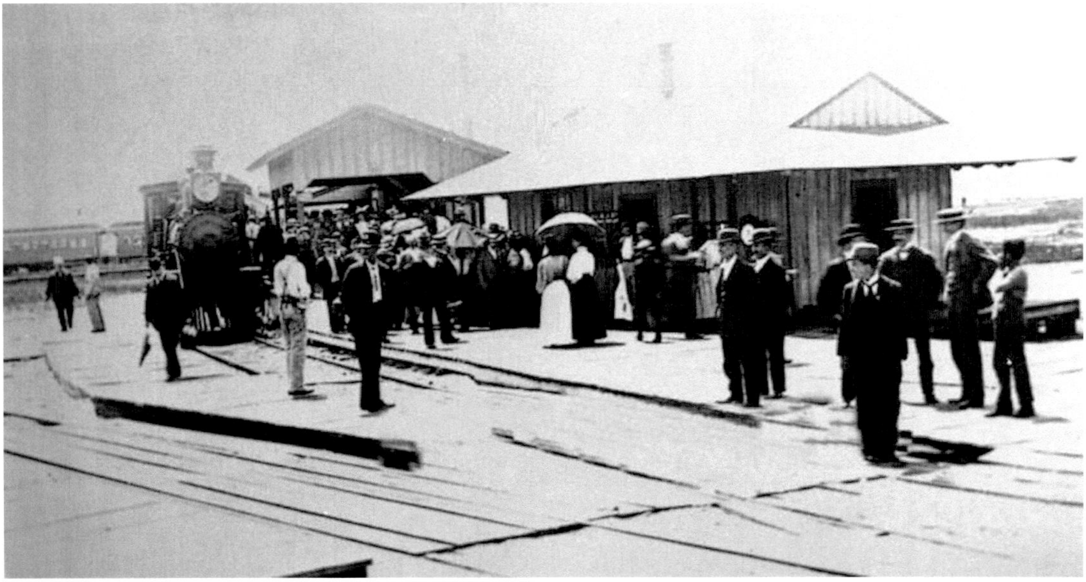

The eastern terminus of the Baltimore, Chesapeake, and Atlantic was Ocean City, Maryland. Originally called "The Ladies' Resort to the Ocean," the area was little more than one hotel and cottages until the railroad began bringing passenger cars filled with tourist from the big cities of the western shore. By 1900, the area had its first boardwalk and amusement pier. Mother Nature however had other plans for the railroad access to the growing resort town. On August 22, 1933, one of the most devastating hurricanes to ever hit Delmarva completely destroyed the entire crossing. [*Courtesy of Jim Bowden & Keith Short*]

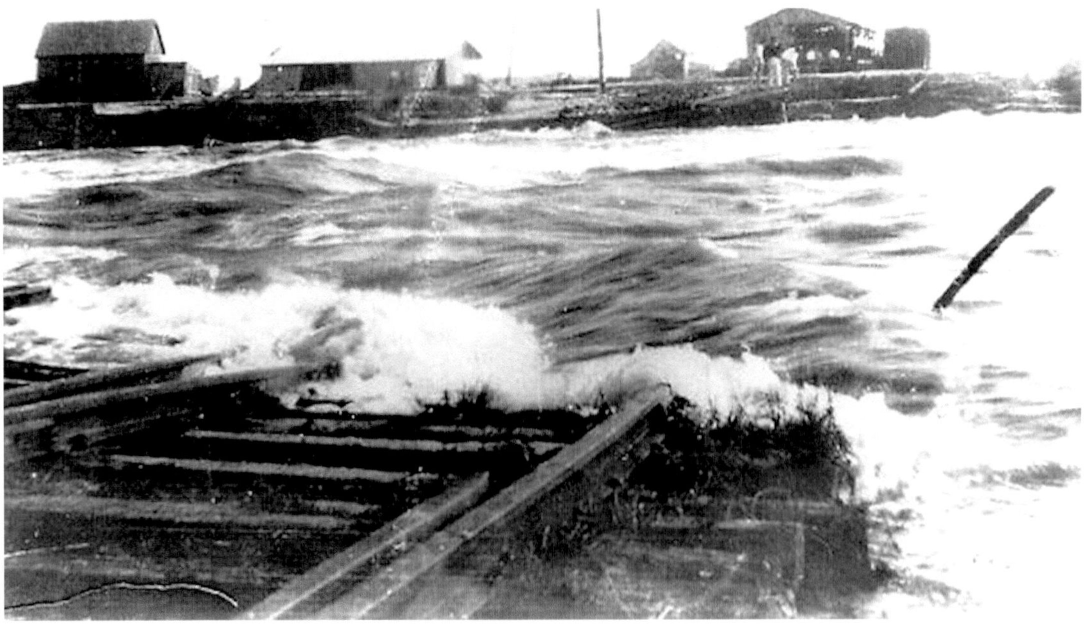

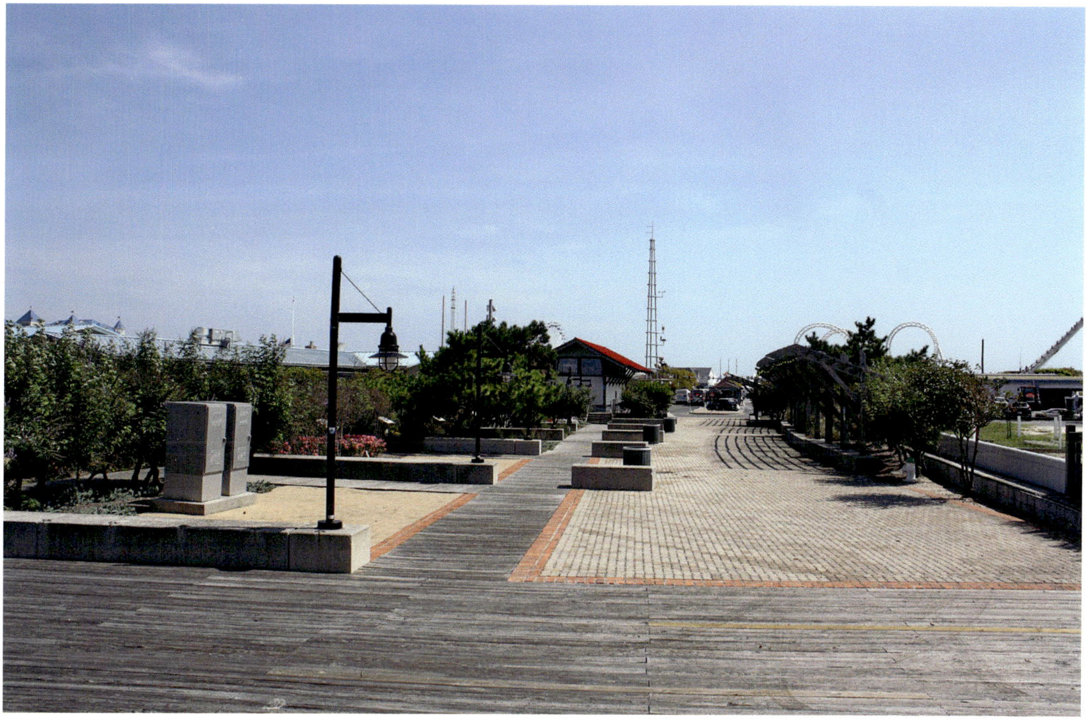

Today, Ocean City pays homage to its railroad past. Constructed as a public parking lot, it has public restrooms and an overlook of Sinepuxent Bay. Visitors at low tide are still able to see a few remains of the train bridge that once spanned the water. The bridge was so crucial to the development of Ocean City that prior to 1916, the few cars that visited Ocean City had to pay a five-cent toll to cross the railroad bridge on wooden planks.

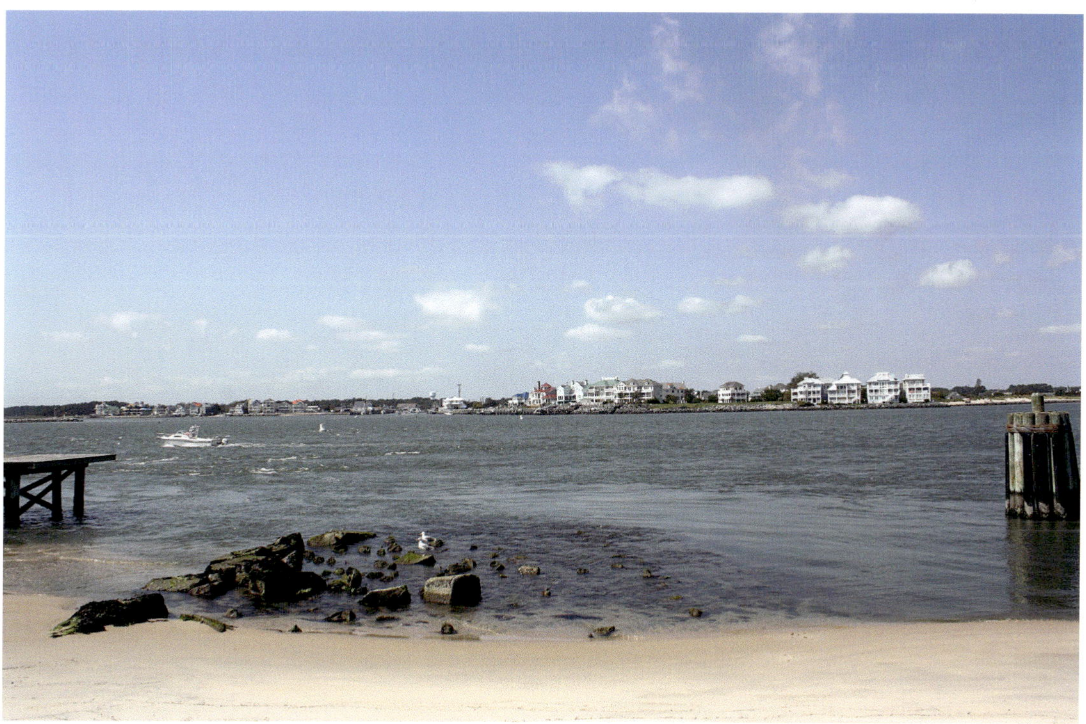

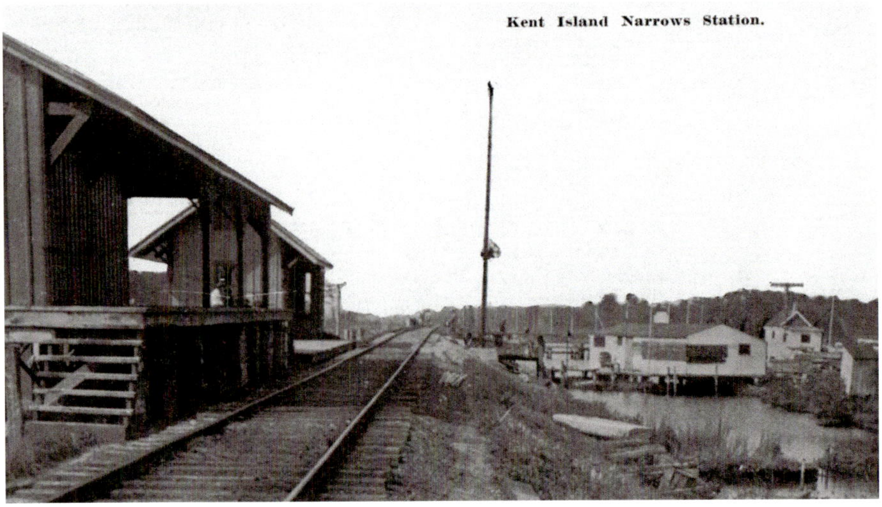

When first constructed in 1897, the Queen Anne's Railroad ferry terminal was located in Kent Island Maryland. The city was the first English settlement in Maryland way back in 1627. The city served at the western end of the line till 1902, when it was extended to shorten the crossing time. The Narrows was once the heart of the seafood industry, with twelve packinghouses in operation along its waters. In its heyday in the latter half of the twentieth century, hundreds of boats arrived daily to deliver their catch to the seafood houses. Today, only two packinghouses remain, but the Narrows still boasts numerous seafood restaurants. The remains of the railroad bridge, dismantled in 1957, can still be seen today. [*Courtesy of the Kent Island Heritage Society*]

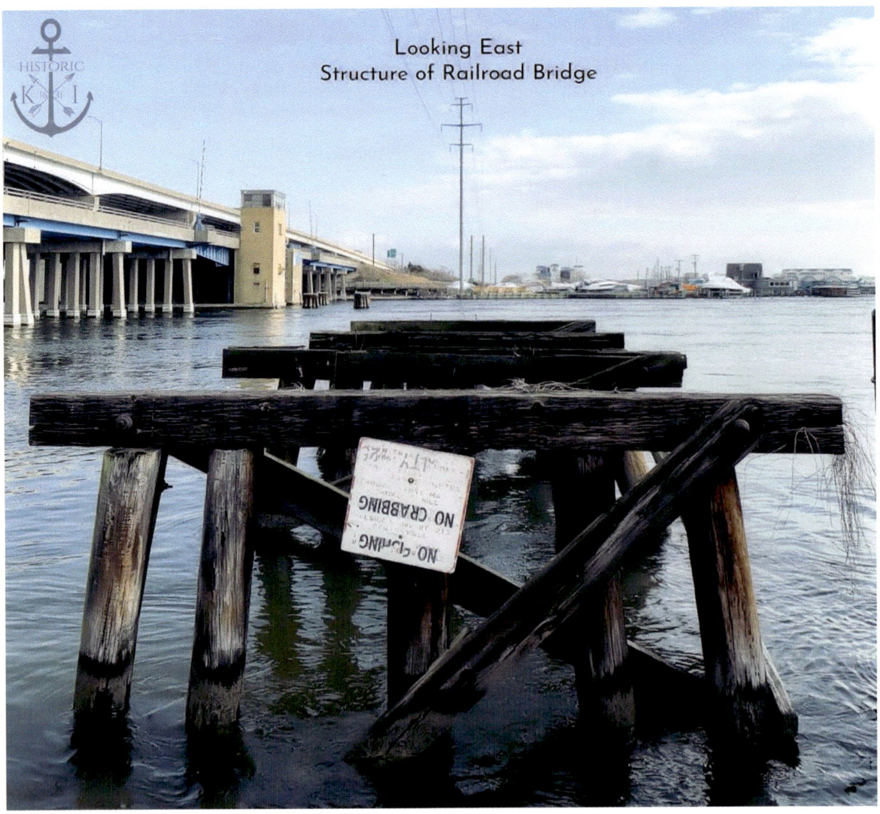

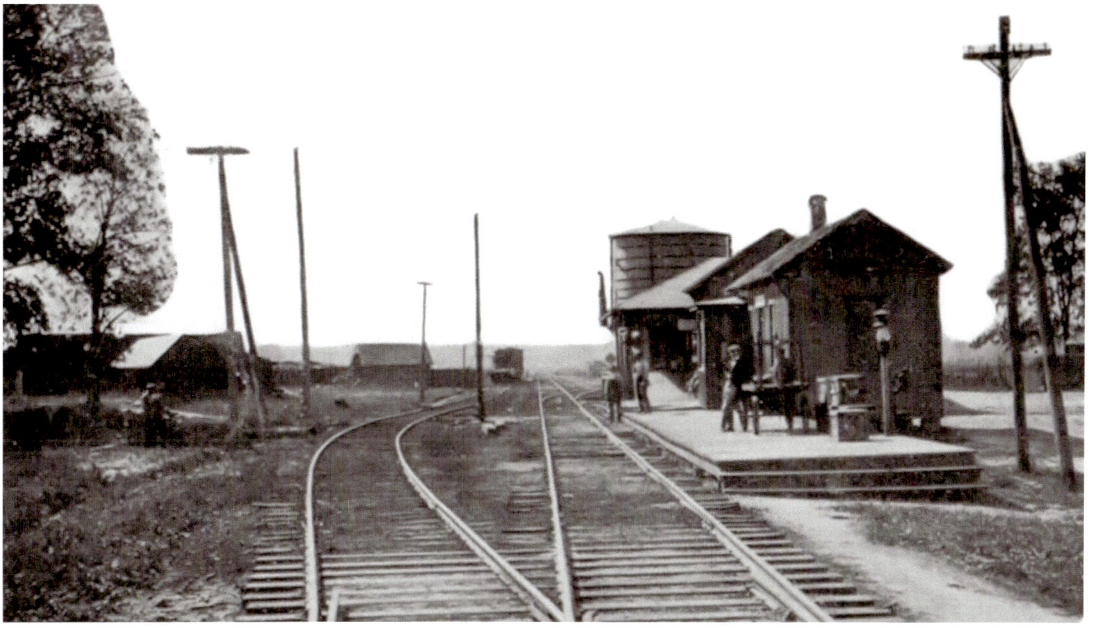

Queenstown, Maryland, was the first of the towns to bear the name "Queen's Town" in 1707. Later the name changed to Queenstown, but it is a mere 19 miles from a previously mentioned place on another railroad, Queen Anne's, Maryland. With both towns having similar names and the railroads sharing the Queen name in their titles, it's sometimes hard to keep both straight. The station here was constructed to connect steamship landings to the railroad.

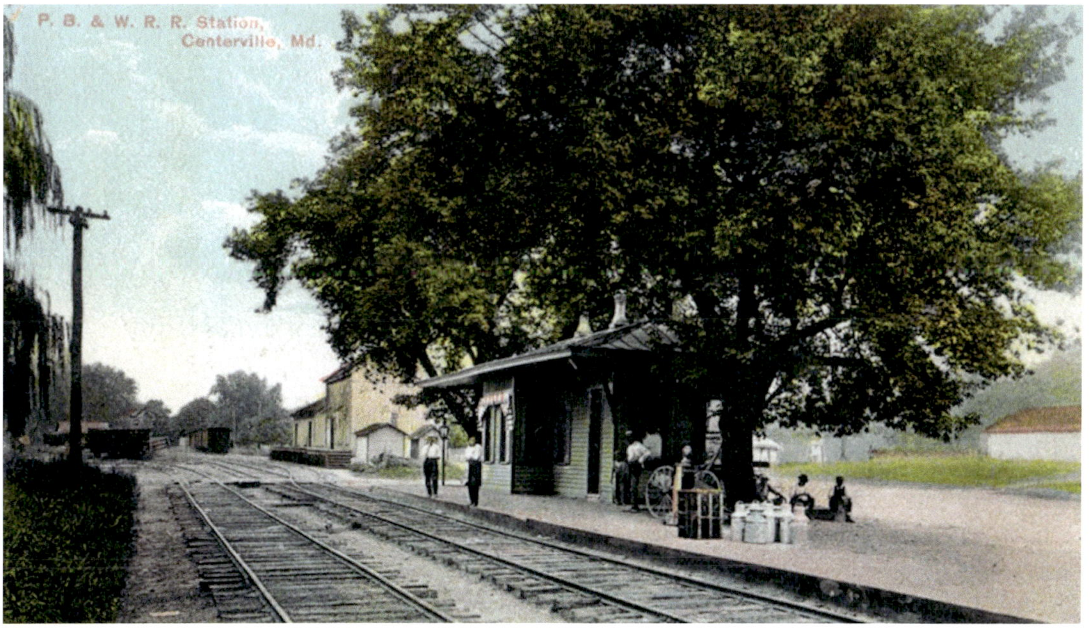

A portion of the Queen Ann's Railroad was called the Centerville Branch and finally achieved the goal of redirecting goods from going north and instead sending them westward to Baltimore and beyond. Centerville, Maryland, was a major wharf landing long before the railroad, and the connection with the railroad only increased its importance.

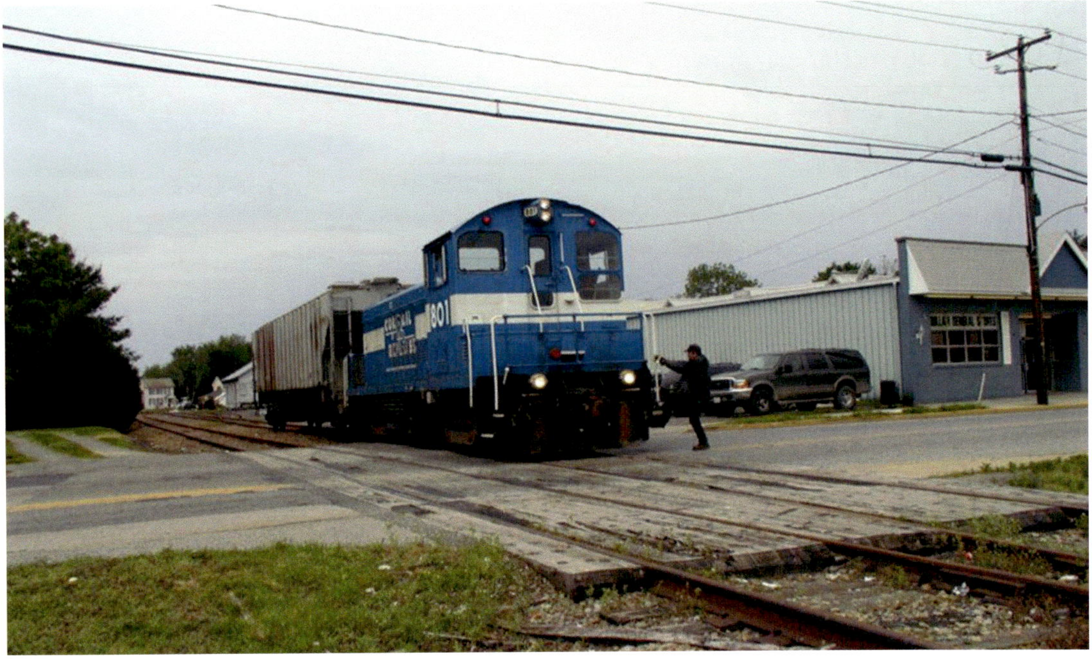

The Centerville Branch continued its importance to the grain industry on Delmarva until 2003, when finally the last train heading west departed the city. Today, the rails sit abandoned in parts of the town while others have been removed. The passenger station is gone, but the freight depot has been moved and there are current plans to restore the building. Centerville still has rail service today on the old Townsend Branch line, a part what once was the Queen Anne and Kent Railroad. Today, the Maryland and Delaware Railroad runs trains from Townsend, Delaware, to the north portion of the town.

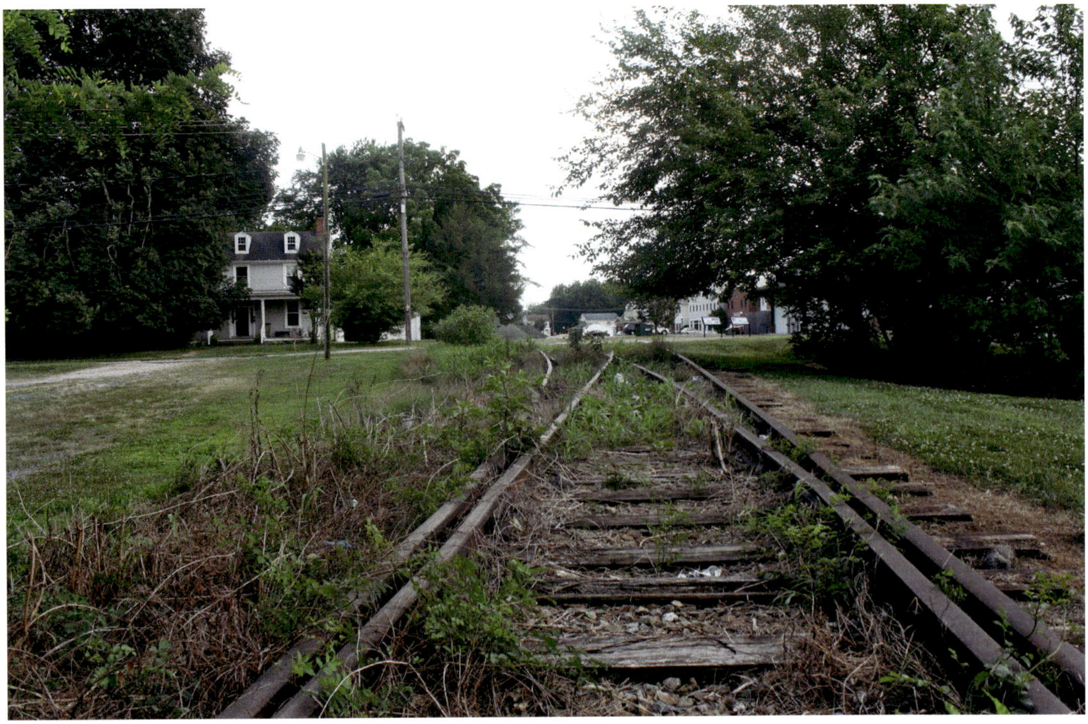

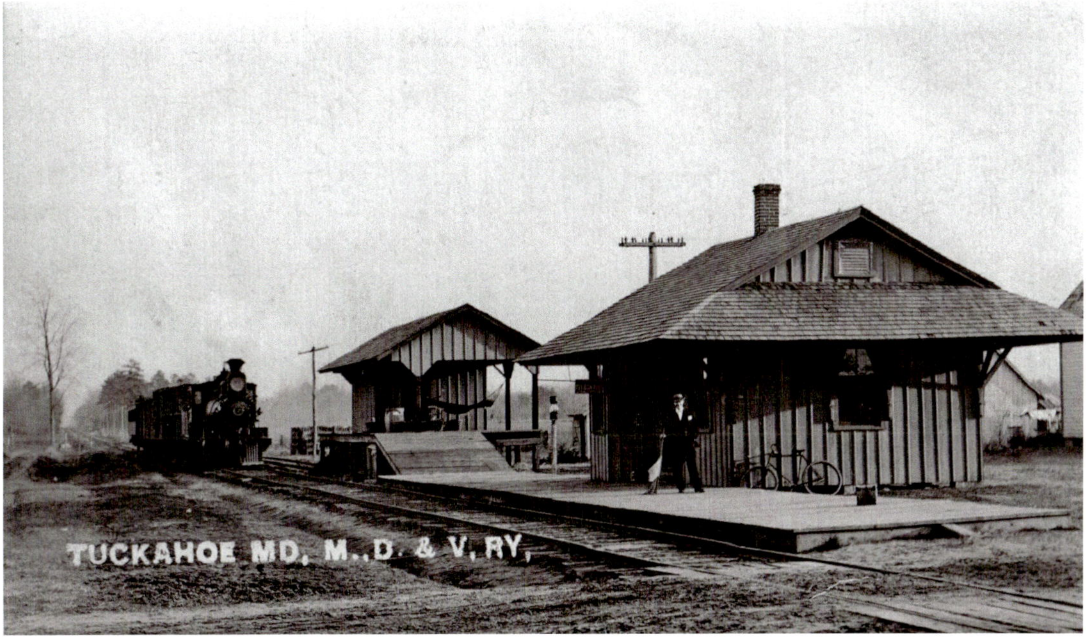

Some stations seem to have truly sprung up from the dirt. Tuckahoe, Maryland, station is a perfect example of this. While the creek that ran close to the station was a long winding waterway that had been used by locals in the area, there were no mills, canneries, or even steamship wharfs in the area. Yet the railroad viewed the area as important enough to receive its own stop on the line. [*Courtesy of Jim Bowden & Keith Short*]

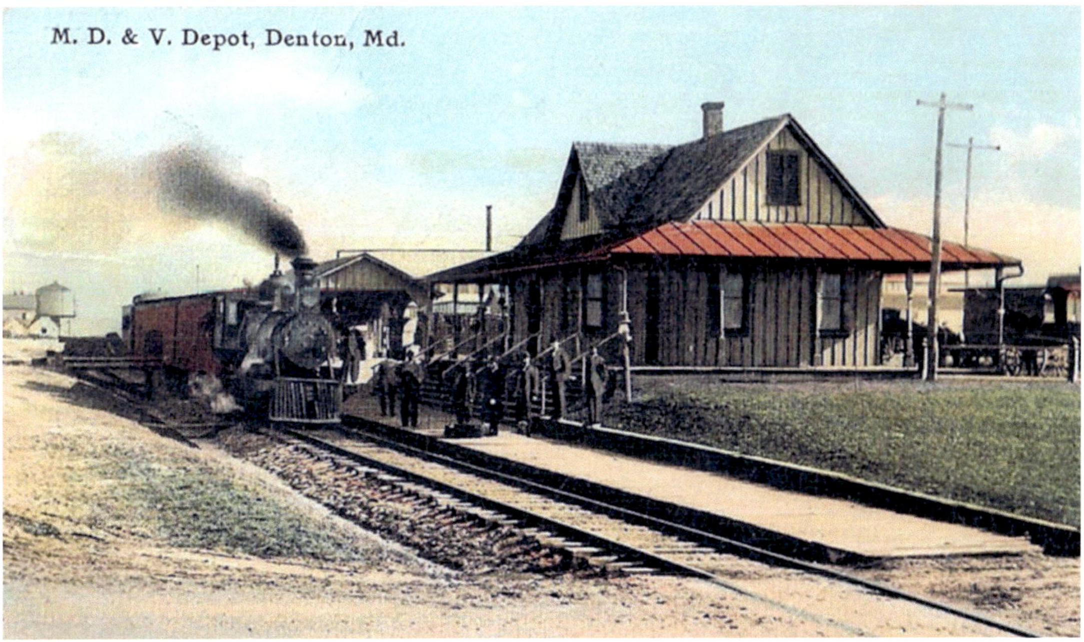

In stark contrast to the previous station, Denton, Maryland, was an obvious choice for a railroad station. Sitting along the Choptank River, Denton early on was established a bustling port which only grew in importance with the steamship. When then railroad arrived in the early 1900s, the population of the town tripled in size within ten years.

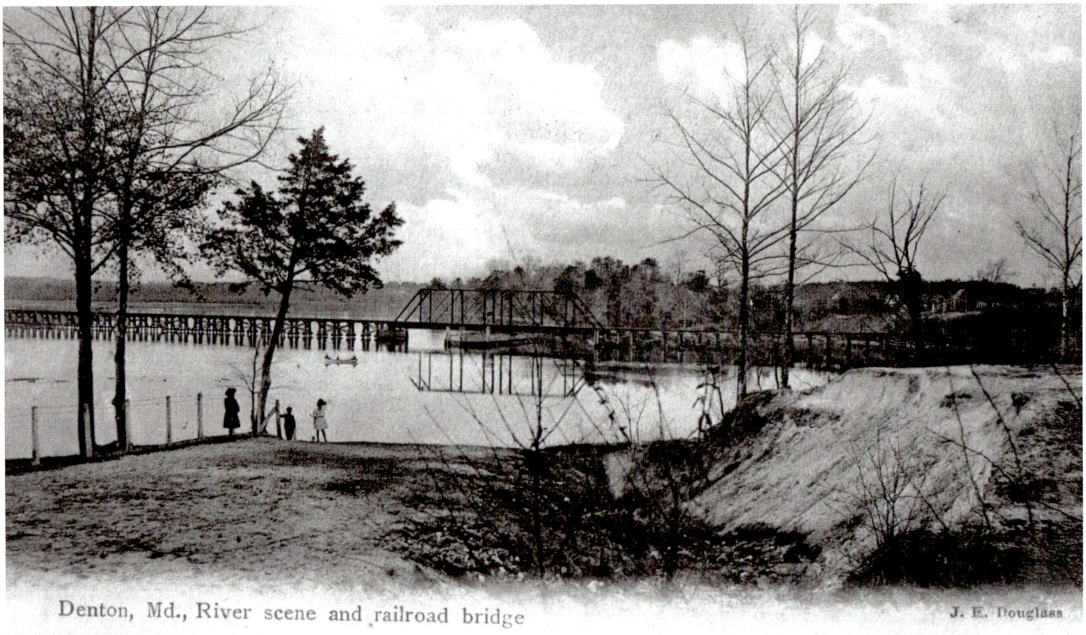

While Denton was destined to be a busy railway station, the engineers faced the challenge of allowing the busy steamship piers along the Choptank to remain uninterrupted, and placed a bridge over the waterway. The tried and true swing bridge was chosen to span across the river. The crossing was too wide for a drawbridge, but by constructing a man-made island in the center of the river, the ships could pass and the trains could roll across the expanse. The wooden trestles were so well designed that only maintenance and minor repair was ever done to them. The old bridge was replaced by a new steel structure at an unknown date. Amazingly enough, despite no maintenance on the bridge since the 1980s, it still stands firmly and seems to be ready for its next crossing. Photo 101 [*Courtesy of Jim Bowden & Keith Short*]

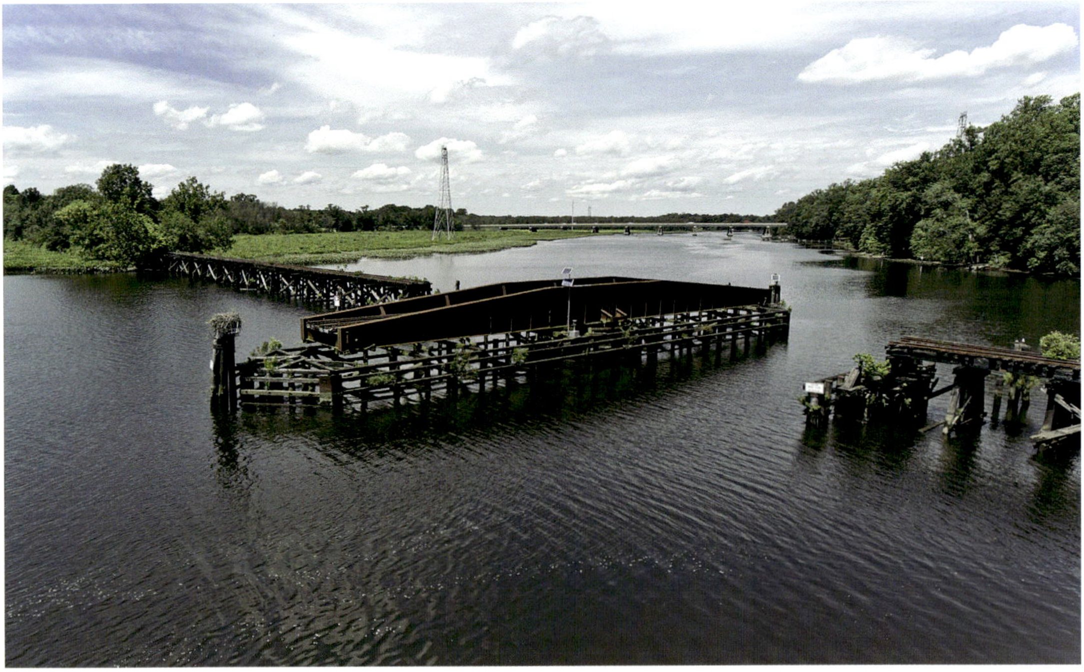

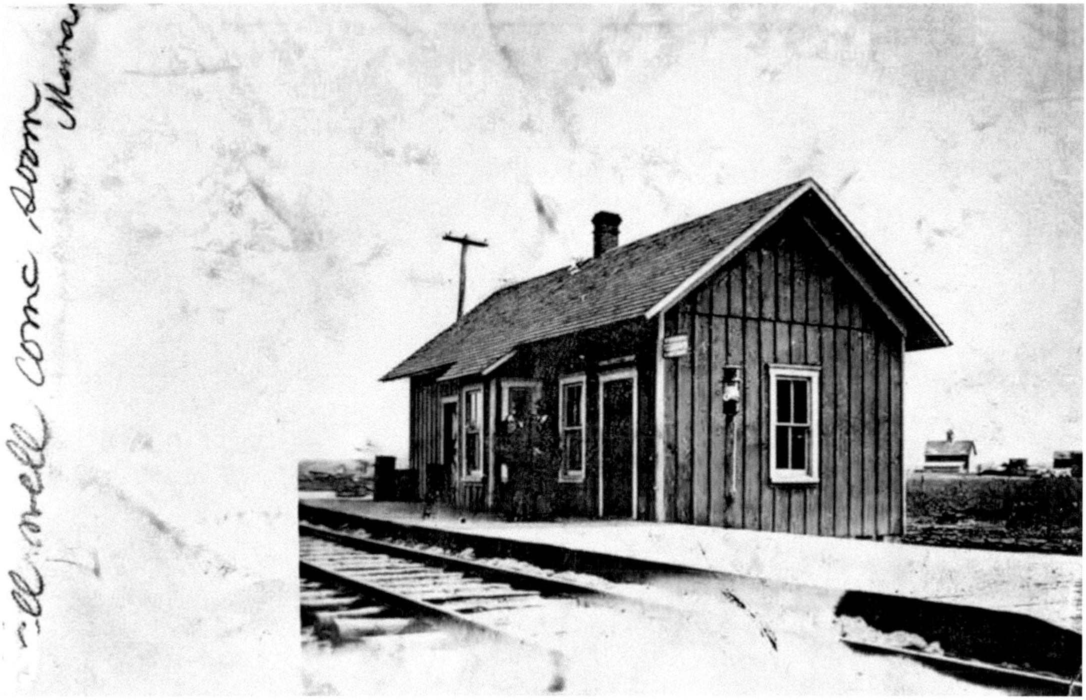

Hickman, Delaware, was and remains a simple crossroads on the western boundary of Delaware with Maryland. Homes have dotted both sides of the roads since the area became a settlement in the early 1800s. When the railroad came though, a small station was placed on the line because of a few farms and canneries in the area. The station also became a frequent meeting spot for the locals since it has the only restaurant and tavern for miles. The area, however, was made "dry" by the local county legislature as prohibition in America began. So, the local residents petitioned the railroad to move the station about the required 300 feet into Maryland so they could resume liquor sales. The railroad agreed, but said the citizens would have to move it. And so they did! [*Photo courtesy of Jim Bowden & Keith Short*]

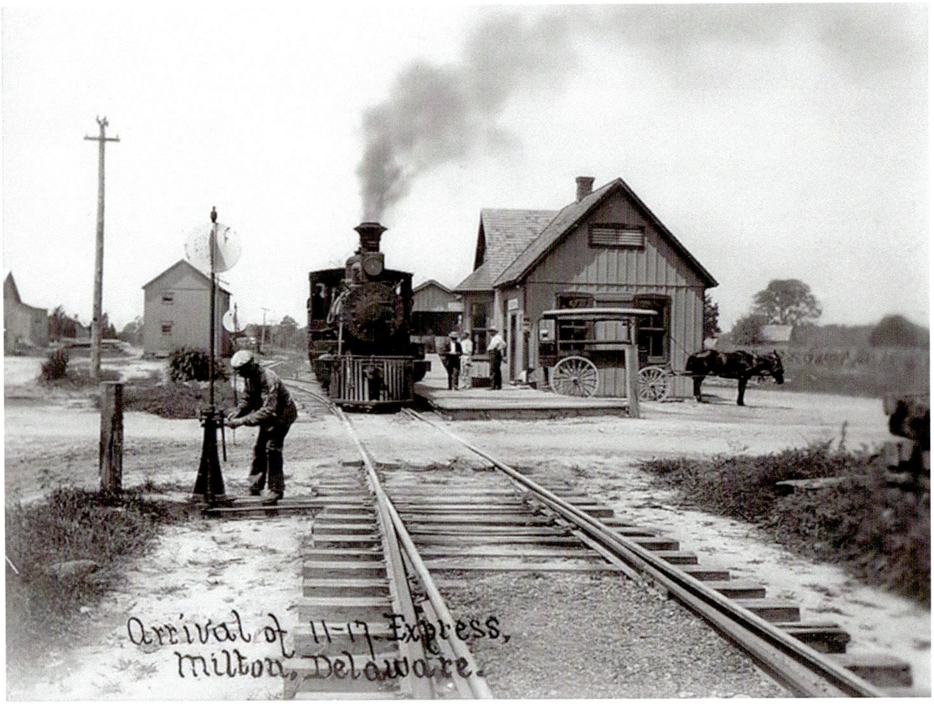

Railroads across America frequently placed depots near areas were shipping was a major industry, and Milton, Delaware, fit that description in every way. The area was settled in 1763 as an inland waterway which gave access to farms, timber, and other goods, and provided safe haven from storms and pirates along the coast. Five to six steamers a day would travel to and from Philadelphia during the railroad's time in Milton. The railroad increased industry in the area and changed it from its inland seaport past to a rail hub that moved freight from seven granaries, multiple lumber mills and even such unique items as holly from the trees during the Christmas season. The Delaware Central Railroad now serves Milton, but the end of the line stops northwest of the city. All remnants of the station and rails in the town were removed and only a bike trail remains. [*Top photo courtesy of Jim Bowden & Keith Short*]

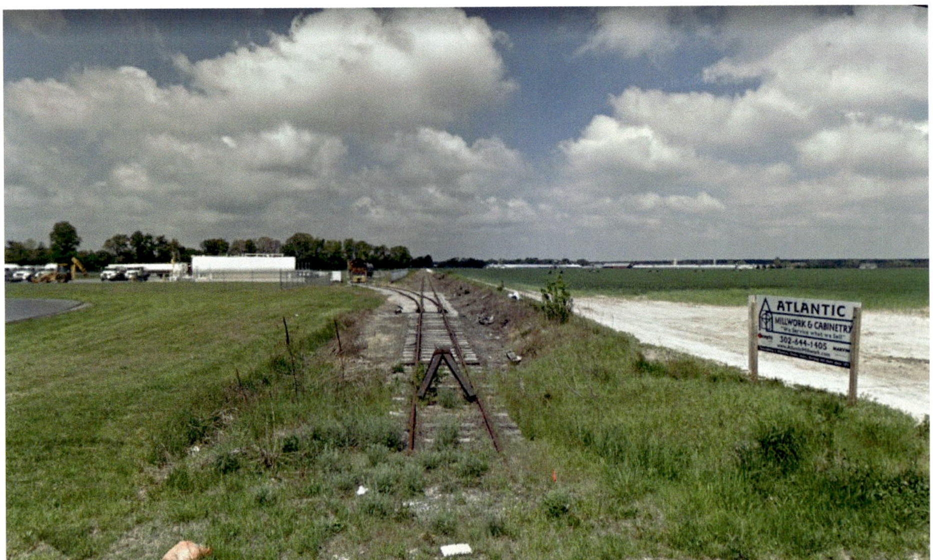

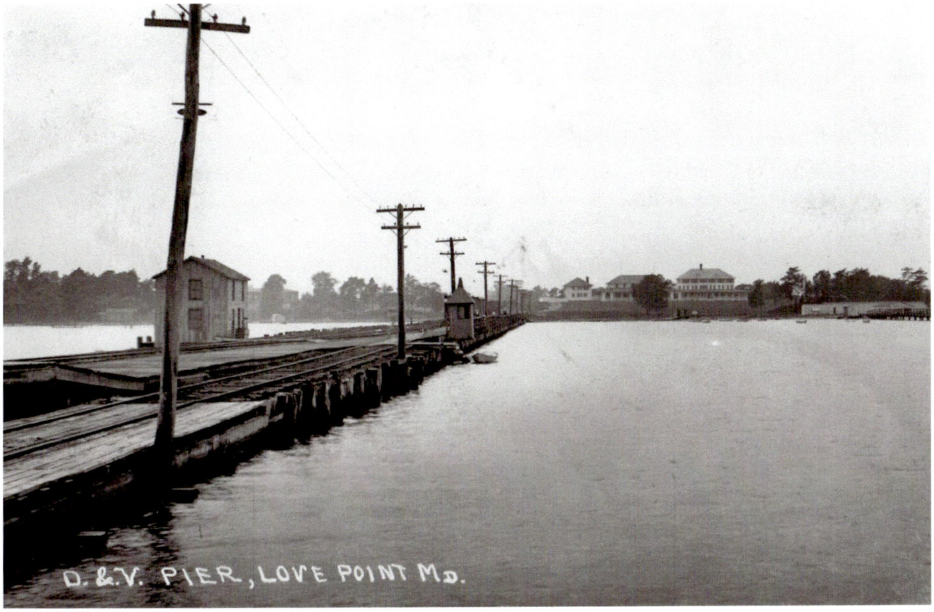

In 1902, the Queen Anne's Railroad decided to shorten the ferry crossing time. This meant a new wharf and all the needed infrastructure required to dock the steamers and move passengers and freight from one to the other. Love Point, Maryland, was chosen to be the new site and the area quickly became a tourist destination. A casino hotel, a sandy beach, expansive lawns, and picnic groves dotted with cottonwood trees, a lake with canoe rentals, an amusement boardwalk with a merry-go-round, a bowling alley, and a shooting gallery all waited for the travelers from the western shore. Passenger service stopped to the area in 1938, the hotel closed in 1947, rail traffic ended in 1958, with the tracks removed, and the hotel burned to the ground in 1964. Rusting metal and weeds bear little resemblance to the grandeur the area once held. [*Top photo courtesy of Kent Island Heritage Society*]

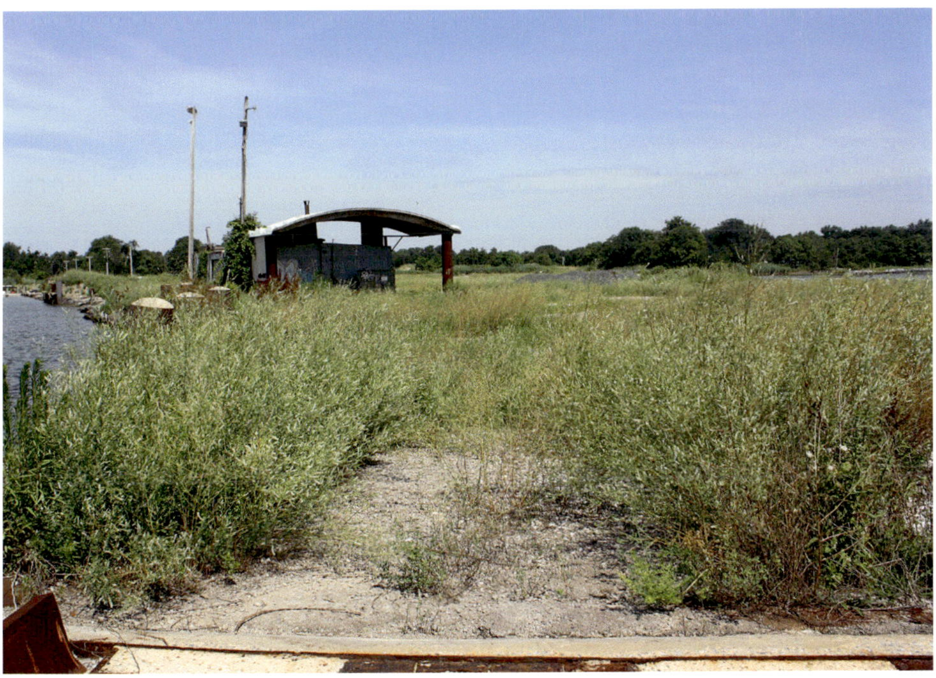

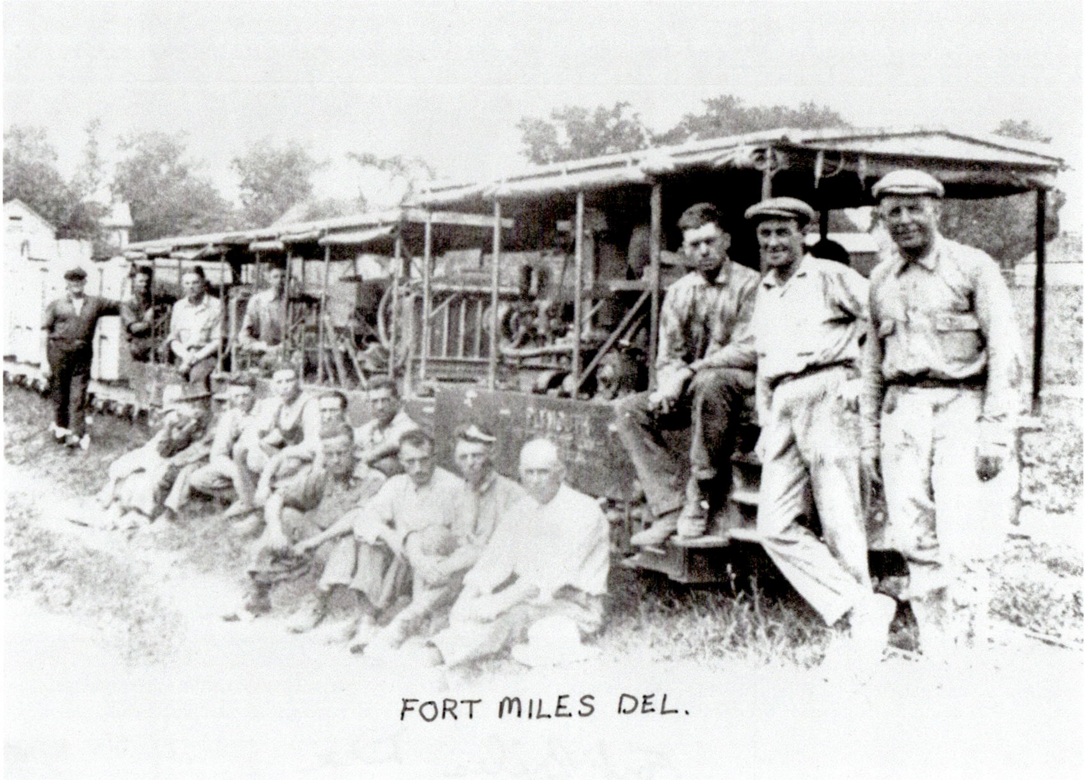
FORT MILES DEL.

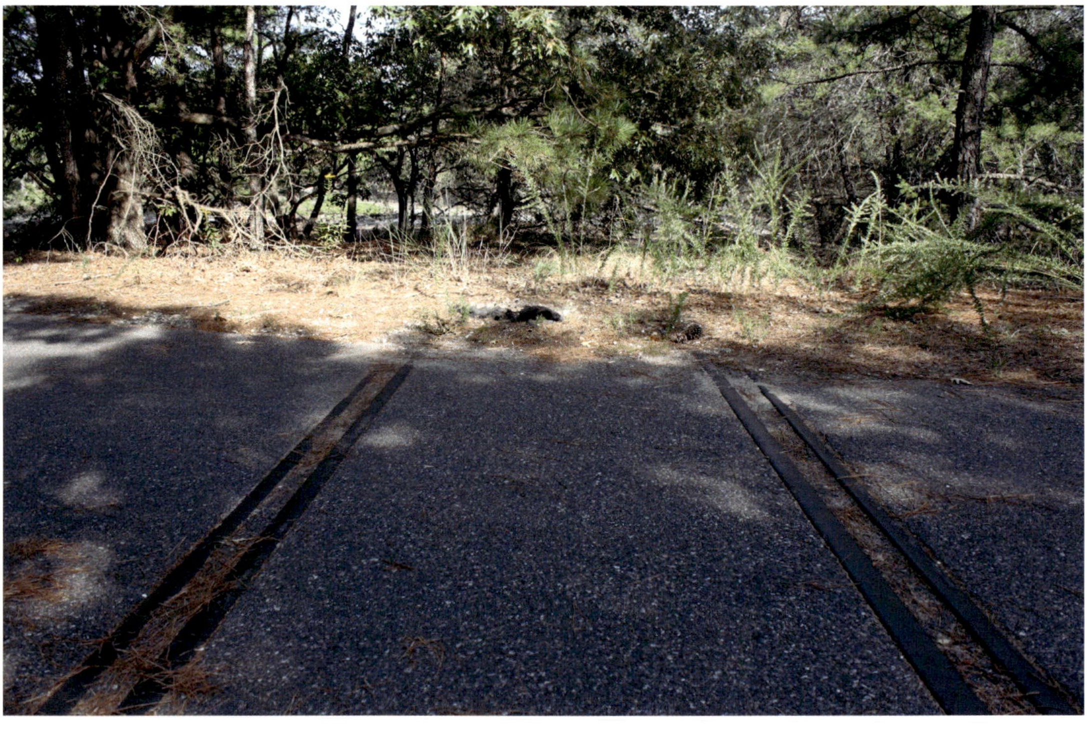

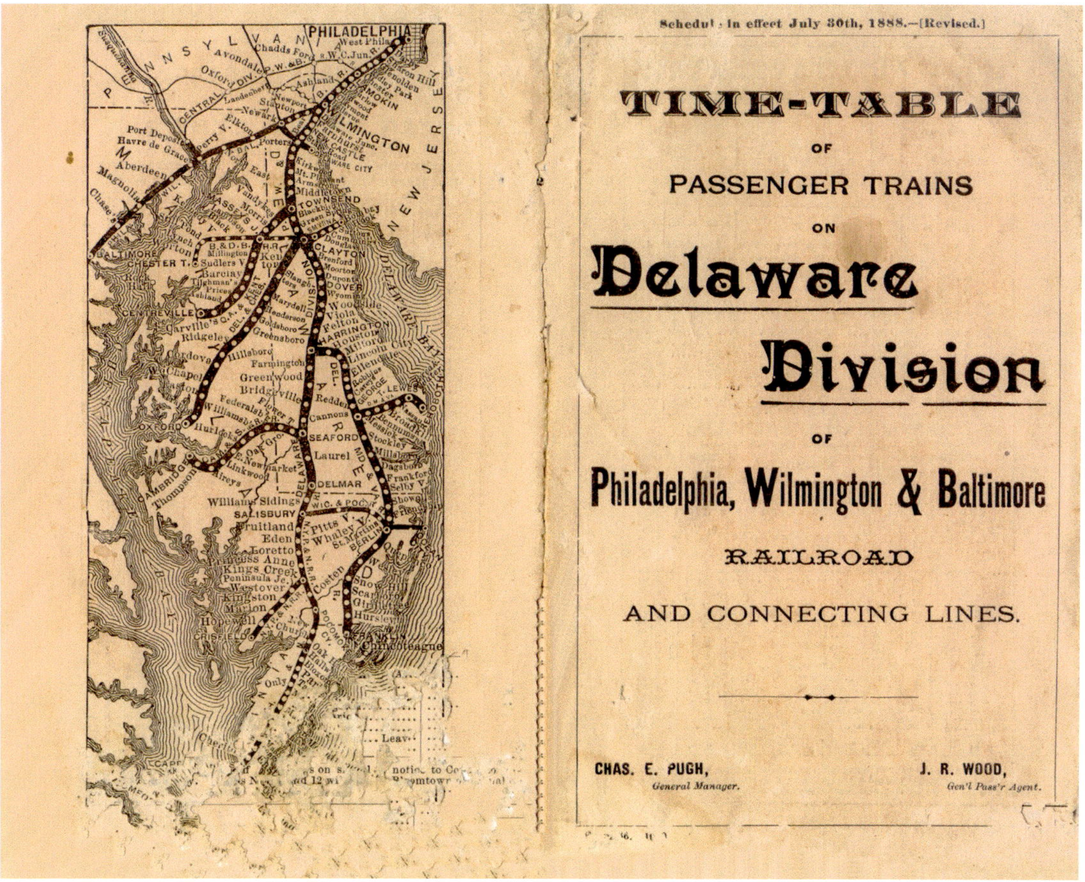

By the 1880s, all the railroads on Delmarva had fallen under the Philadelphia, Wilmington and Baltimore flag. While each line maintained its own name, behind the scenes each group was being controlled by the PW&B. What not many knew outside of the railroads was that the real company running these lines was the mighty Pennsylvania Railroad. Each line's train schedule would coincide with the others to ensure passenger and freight movements were seamless operations. Ultimately these trains would arrive in one of the hub cities of the PRR, thus allowing the rich harvest of Delmarva's land and sea to feed the hungry cities. A businessman could now travel in one day from all the points on the line, conduct their business and be home that night to sleep in their own bed. Consolidating the lines also meant each could share the other's stations and equipment. [*Photo courtesy of Jim Bowden & Keith Short*]

On the previous page: At the eastern end of the Queen Anne's Railroad was Lewes, Delaware. Fisheries were the main draw, particularly a fish called menhaden, which is an oily herring-type fish that is used primarily for fertilizer and animal feed. However, a more important use for the railroad in this area came about with Fort Miles. As the primary fort of the Harbor Defenses of the Delaware, it was built to defend Delaware Bay and the Delaware River. On March 14, 1942, Battery C of the 52nd Coast Artillery Railway regiment arrived with four 8-inch railway guns. At its peak, Fort Miles was home to over 2,200 soldiers. In April 1944 the 8-inch railway gun battalion was relocated to Fort Bragg, North Carolina. The Army continued to use portions of it through the early 1990s as a morale, welfare and recreation area for active and retired military personnel and their families. In 1964, 543 acres of federal land were donated to the State of Delaware to establish Cape Henlopen State Park. Its last official usage was as a bivouac for soldiers who had just returned from the first Gulf War. The only remaining rail lines are embedded in the asphalt near the campgrounds.

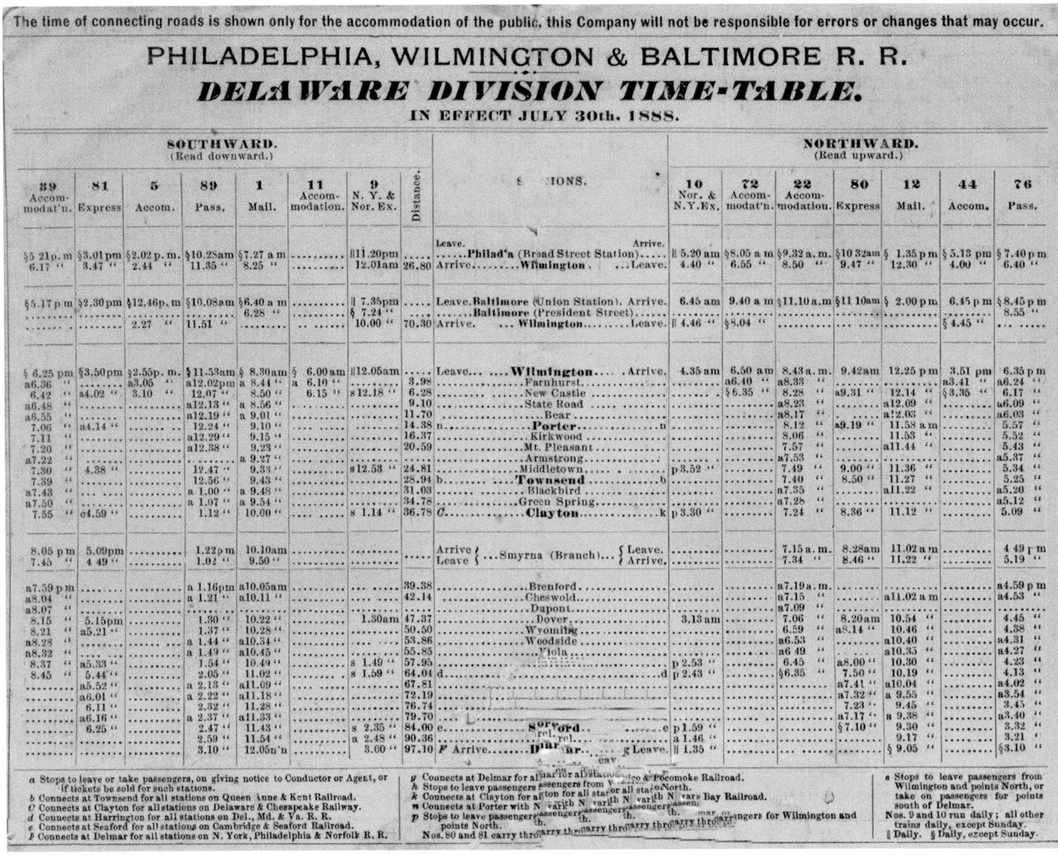

The time of connecting roads is shown only for the accommodation of the public, this Company will not be responsible for errors or changes that may occur.

PHILADELPHIA, WILMINGTON & BALTIMORE R. R.

DELAWARE DIVISION TIME-TABLE.

IN EFFECT JULY 30th, 1888.

SOUTHWARD. (Read downward.)									NORTHWARD. (Read upward.)						
39 Accom-modat'n.	81 Express	5 Accom.	89 Pass.	1 Mail.	11 Accom-modation.	9 N. Y. & Nor. Ex.	Distance.	STATIONS.	10 Nor. & N.Y.Ex.	72 Accom-modat'n.	22 Accom-modation.	80 Express	12 Mail.	44 Accom.	76 Pass.
§5 21p.m	§3.01pm	§2.02 p.m.	§10.28am	§7.27 a m		‖11.20pm		Leave. ......Philad'a (Broad Street Station)..... Arrive.	‖ 5.20 am	§8.05 a m	§9.32 a.m.	§10 32am	§ 1.35 p m	§ 5.13 pm	§ 7.40 p m
6.17 "	3.47 "	2.44 "	11.35 "	8.25 "		12.01am	26.80	Arrive........Wilmington. ...Leave.	4.40 "	6.55 "	8.50 "	9.47 "	12.30 "	4.00 "	6.40 "
§5.17 p m	§2.30pm	§12.46p.m	§10.08am	§6.40 a m		‖ 7.35pm		Leave. Baltimore (Union Station). Arrive.	6.45 am	9.40 a m	§11.10 a.m	§11 10am	§ 2.00 p m	6.45 p m	§8.45 p m
				6.28 "		§ 7.24 "		.......Baltimore (President Street)......							8.55 "
		2.27 "	11.51 "			10.00 "	70.30	Arrive. ... Wilmington........Leave.	‖ 4.46 "	§8.04 "				§ 4.45 "	
§ 6.25 pm	§3.50pm	§2.55p.m.	§11.53am	§ 8.30am	§ 6.00am	‖12.05am		Leave... ....Wilmington.... .Arrive.	4.35 am	6.50 am	8.43 a.m.	9.42am	12.25 p m	3.51 pm	6.35 p m
a6.36 "		a3.05 "	a12.02pm	a 8.44 "	a 6.10 "		3.98	..........Farnhurst..........		a6.40 "	a8.33 "			a3.41 "	a6.24 "
6.42 "	a4.02 "	3.10 "	12.07 "	8.50 "	6.15 "	s12.18 "	6.28	..........New Castle..........		§6.35 "	8.28 "	a9.31 "	12.14 "	§3.35 "	6.17 "
a6.48 "			a12.13 "	a 8.56 "			9.10	..........State Road..........			a8.23 "		a12.09 "		a6.09 "
a6.55 "			a12.19 "	a 9.01 "			11.70	..........Bear..........			a8.17 "		a12.03 "		a6.03 "
7.06 "	a4.14 "		12.24 "	9.10 "			14.38	n..........Porter..........n			8.12 "	a9.19 "	11.58 a m		5.57 "
7.11 "			a12.29 "	9.15 "			16.37	..........Kirkwood..........			8.06 "		11.53 "		5.52 "
7.20 "			a12.38 "	9.23 "			20.59	..........Mt. Pleasant..........			7.57 "		a11.44 "		5.43 "
a7.22 "				a 9.27 "				..........Armstrong..........			a7.53 "				a5.37 "
7.30 "	4.38 "		12.47 "	9.33 "		s12.53 "	24.81	..........Middletown..........	p3.52 "		7.49 "	9.00 "	11.36 "		5.34 "
7.39 "			12.56 "	9.43 "			28.94	b..........Townsend..........b			7.40 "	8.50 "	11.27 "		5.25 "
a7.43 "			a 1.00 "	a 9.48 "			31.03	..........Blackbird..........			a7.35 "		a11.22 "		a5.20 "
a7.50 "			a 1.07 "	a 9.54 "			34.78	..........Green Spring..........			a7.28 "				a5.12 "
7.55 "	c4.59 "		1.12 "	10.00 "		s 1.14 "	36.78	C..........Clayton..........k	p 3.30 "		7.24 "	8.36 "	11.12 "		5.09 "
8.05 p m	5.09pm		1.22 p m	10.10am				Arrive ...Smyrna (Branch)... Leave.			7.15 a.m.	8.28am	11.02 a m		4 49 p m
7.45 "	4 49 "		1.02 "	9.50 "				Leave ... Arrive.			7.34 "	8.46 "	11.22 "		5.19 "
a7.59 p m			a 1.16pm	a10.05am			39.38	..........Brenford..........			a7.19 a.m.				a4.59 p m
a8.04 "			a 1.21 "	a10.11 "			42.14	..........Cheswold..........			a7.15 "		a11.02 a m		a4.53 "
a8.07 "								..........Dupont..........			a7.09 "				
8.15 "	5.15pm		1.30 "	10.22 "		1.30am	47.37	..........Dover..........	3.13 am		7.06 "	8.20am	10.54 "		4.45 "
8.21 "	a5.21 "		1.37 "	10.28 "			50.50	..........Wyoming..........			6.59 "	a8.14 "	10.46 "		4.38 "
a8.28 "			a 1.44 "	a10.34 "			53.86	..........Woodside..........			a6.53 "		a10.40 "		a4.31 "
a8.32 "			a 1.49 "	a10.45 "			55.85	..........Viola..........			a6 49 "		a10.35 "		a4.27 "
8.37 "	a5.33 "		1.54 "	10.49 "		s 1.49 "	57.95	[illegible]	p 2.53 "		6.45 "	a8.00 "	10.30 "		4.23 "
8.45 "	5.44 "		2.05 "	11.02 "		s 1.59 "	64.04	d..........d	p 2.43 "		§6.35 "	7.50 "	10.19 "		4.13 "
	a5.52 "		a 2.13 "	a11.09 "			67.81					a7.41 "	a10.04 "		a4.02 "
	a6.01 "		a 2.22 "	a11.18 "			72.19					a7.32 "	a 9.55 "		a3.54 "
	6.11 "		2.32 "	11.28 "			76.74					7.23 "	9.45 "		3.45 "
	a6.16 "		a 2.37 "	a11.33 "			79.70					a7.17 "	a 9.38 "		a3.40 "
	6.25 "		2.47 "	11.43 "		s 2.35 "	84.00	e..........[illegible]ford..........e	p1.59 "			§7.10 "	9.30 "		3.32 "
			2.59 "	11.54 "		a 2.48 "	90.36	[illegible]	a 1.46 "				9.17 "		3.21 "
			3.10 "	12.05n'n		3.00 "	97.10	F Arrive.......[illegible]........g Leave.	‖ 1.35 "				§ 9.05 "		§3.10 "

a Stops to leave or take passengers, on giving notice to Conductor or Agent, or if tickets be sold for such stations.
b Connects at Townsend for all stations on Queen Anne & Kent Railroad.
C Connects at Clayton for all stations on Delaware & Chesapeake Railway.
d Connects at Harrington for all stations on Del., Md. & Va. R. R.
e Connects at Seaford for all stations on Cambridge & Seaford Railroad.
F Connects at Delmar for all stations on N. York, Philadelphia & Norfolk R. R.

g Connects at Delmar for al[illegible] & Pocomoke Railroad.
h Stops to leave passengers [illegible] North.
k Connects at Clayton for a[illegible] Bay Railroad.
n Connects at Porter with N[illegible]
p Stops to leave passengers [illegible] for Wilmington and points North.
Nos. 80 and 81 carry thro[illegible]

s Stops to leave passengers from Wilmington and points North, or take on passengers for points south of Delmar.
Nos. 9 and 10 run daily; all other trains daily, except Sunday.
‖ Daily. § Daily, except Sunday.

The backbone of the PW&B was now known as the Delaware Division, the original Delaware Railroad. Seven passenger trains a day went north and south. Some of these would stop at every station while others were express trains, only stopping at busier locations. Connections to the smaller lines were available three times a day, one of which included a mail car serving each location. The express ran from Philadelphia to Delmar in just over three hours. Then a rider could transfer over to the New York, Philadelphia, and Norfolk line and continue their journey on to the port city of Norfolk, Virginia. A trip that use to take a week or more now allowed the traveler to complete the journey in twelve hours. But it was not just about point-to-point travel; it was about opening up over 6,000 square miles of land to products, services, and a way of life never before imagined to those that had lived an isolated existence prior to the railroad. *[Photos courtesy of Jim Bowden & Keith Short]*

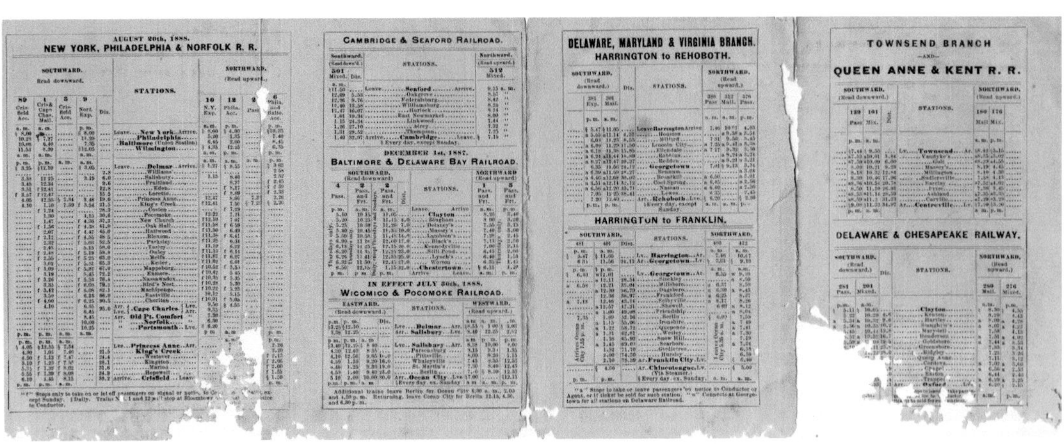

NEW YORK, PHILADELPHIA & NORFOLK R. R.

CAMBRIDGE & SEAFORD RAILROAD.

BALTIMORE & DELAWARE BAY RAILROAD.

WICOMICO & POCOMOKE RAILROAD.

DELAWARE, MARYLAND & VIRGINIA BRANCH.

HARRINGTON to REHOBOTH.

HARRINGTON to FRANKLIN.

TOWNSEND BRANCH

QUEEN ANNE & KENT R. R.

DELAWARE & CHESAPEAKE RAILWAY.

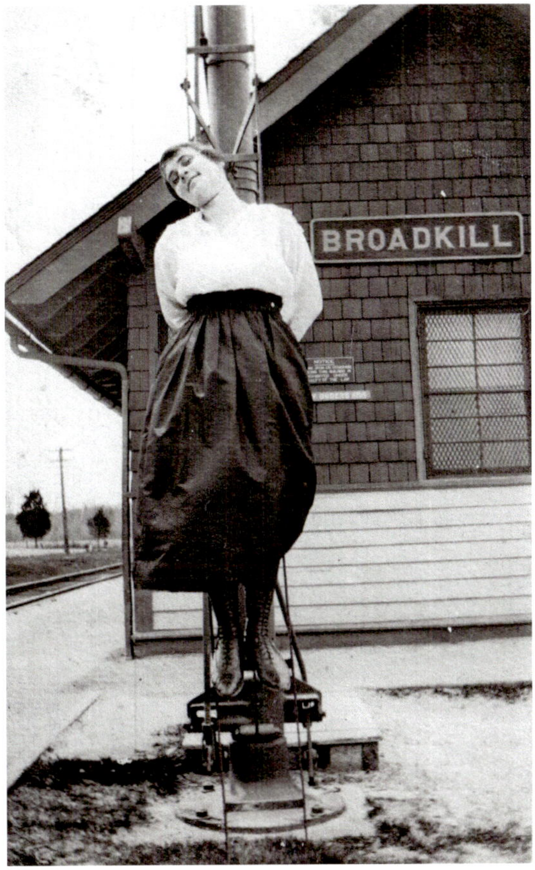

While the railroad created towns all over the Delmarva Peninsula, other stations just borrowed names from towns close by until they eventually grew large enough to gain their own identity. Broadkill Station was 12 miles from Broadkill Beach, which was founded in the early 1700s. When a station was built on a site with no name, the railroad chose to use the name of this nearby beach community. As the area around the station grew, the station was renamed Harbeson, taking the name of a major landowner in the area, Harbeson Hickman. The railroad bed is still in the area, but stops before the former location of the station.

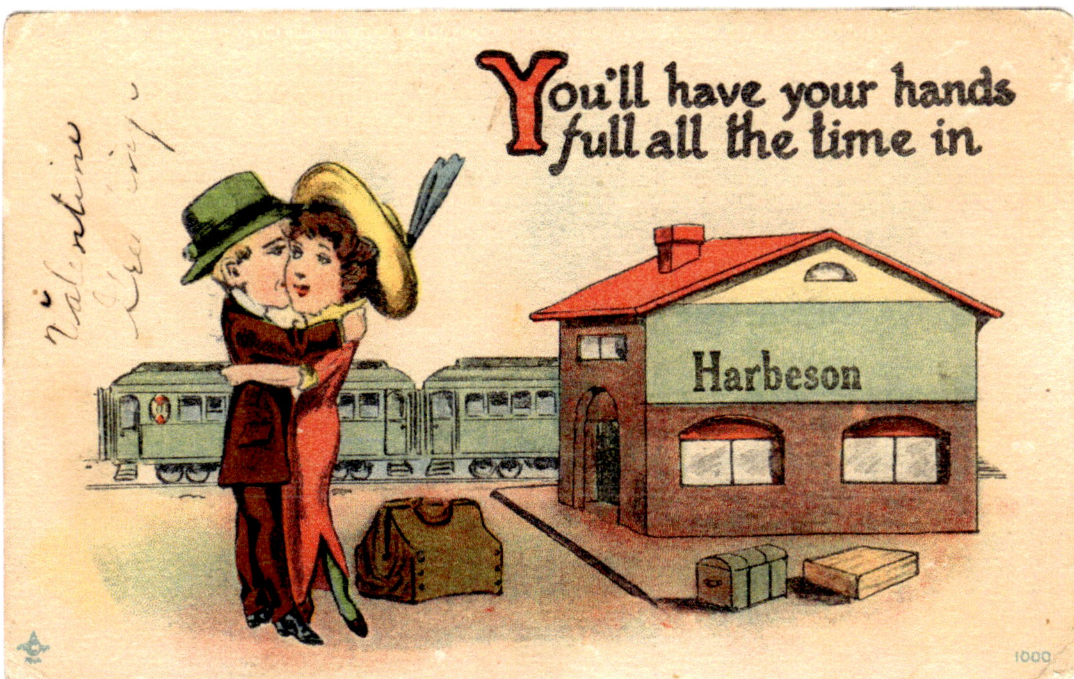

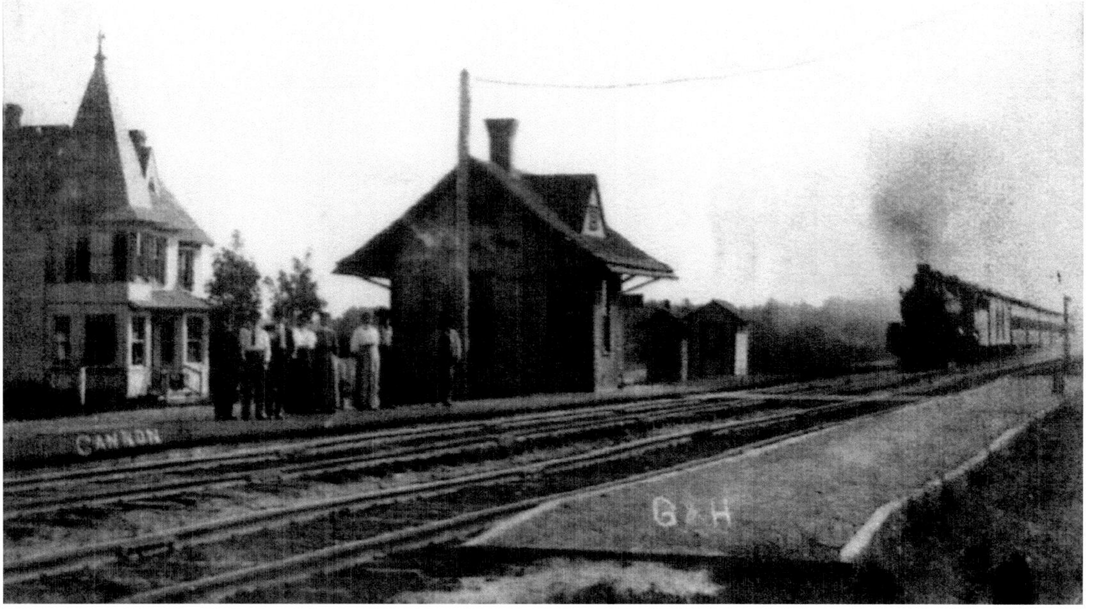

Another town that got a stop on the growing railroad was Cannon, Delaware. As with Harbeson, Cannon was a simple crossroads with a few homes but still had enough freight traffic to earn a stop along the rail lines. The single-room station doubled as a freight station and post office for the area. The area remains virtually unchanged to this day, but the train still passes through the country crossroads. [*Photo courtesy of Jim Bowden & Keith Short*]

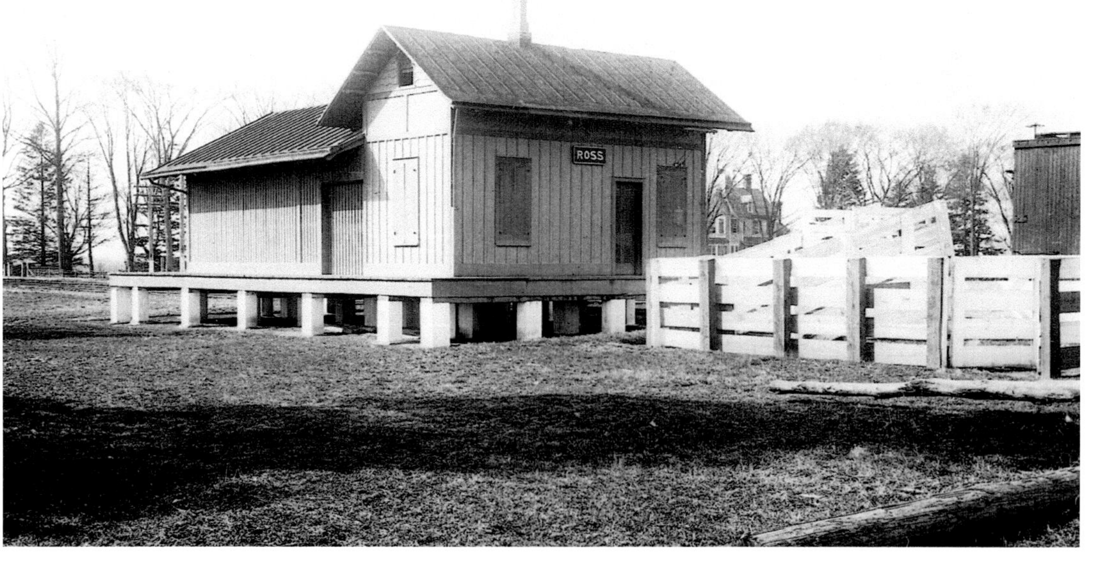

As with such towns as Felton, Clayton, and Harrington, Ross, Delaware, was named after the man who aided the railroad's development—in this case, William Henry Harrison Ross. Ross constructed a large mansion in the area in the 1850s and served as the Governor of Delaware from 1851 to 1855. Seen in this photo is something that was common to every station on Delmarva: a cow pen for the shipping of livestock. [*Photo courtesy of Jim Bowden & Keith Short*]

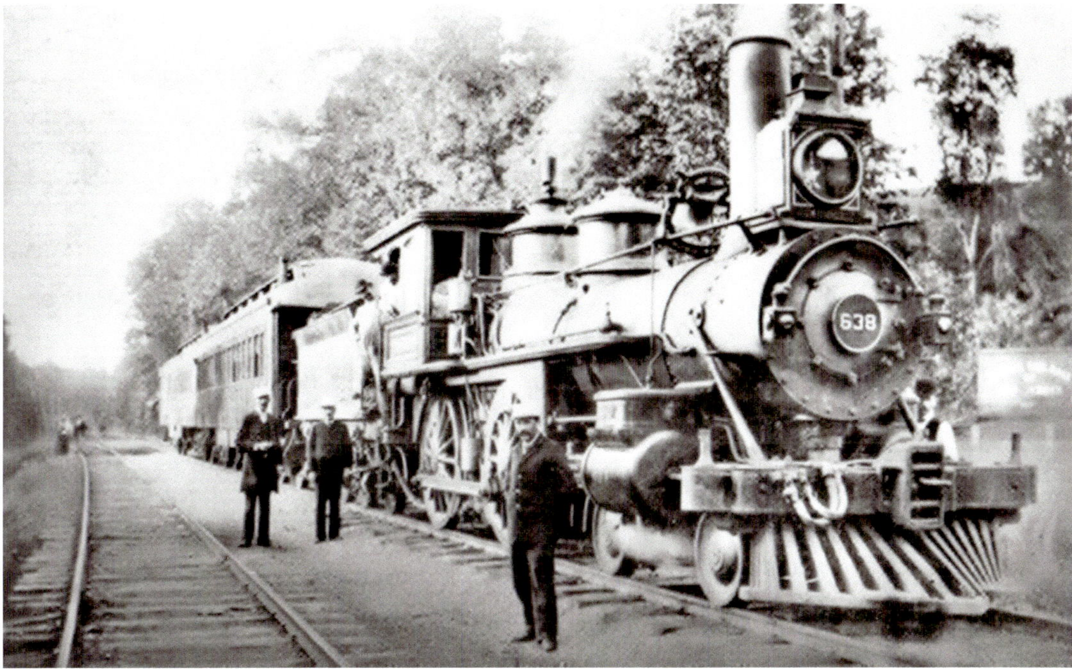

So many areas of Delmarva clamored for the railroad to serve their area. The Wilmington and Western Railroad is by every definition a perfect example of this. A mere 20 miles long, it ran through the Red Clay Creek of New Castle County, Delaware, an area full of water powered mills and farms. The line began as the Delaware and Chester County Railroad, then two years later became the Wilmington Western. The line did have one unique feature: it had a junction with both the Baltimore & Ohio to the north and the PW&B to the south. Landenberg, Pennsylvania, was the northern end of the line. The line now ends just outside of Hockessin Delaware and is known as one of the best tourist railroads in America.

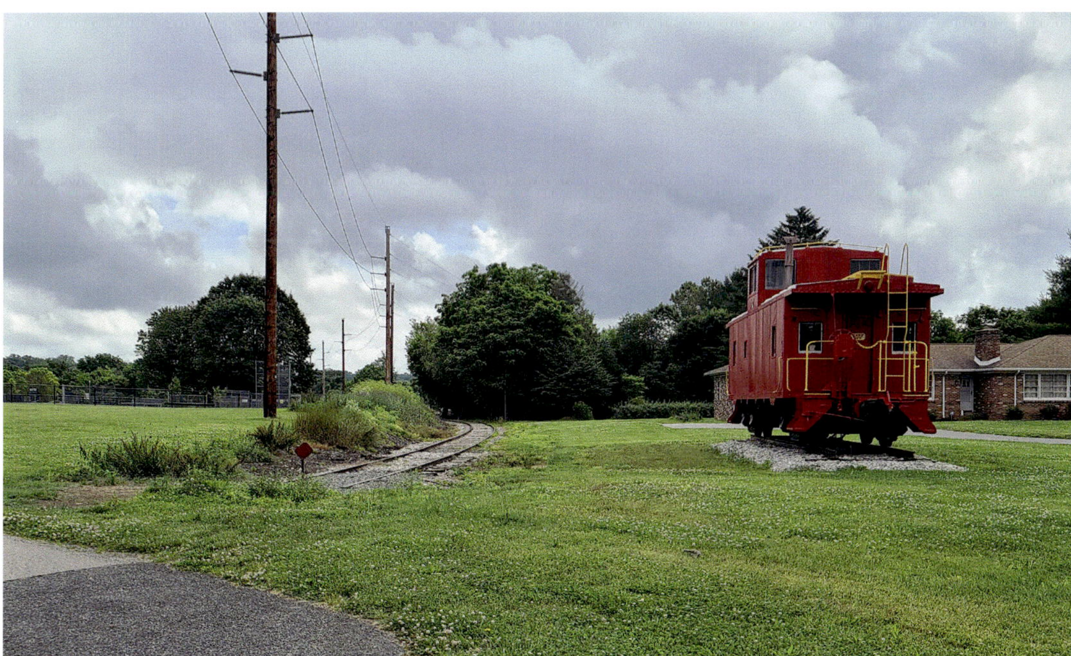

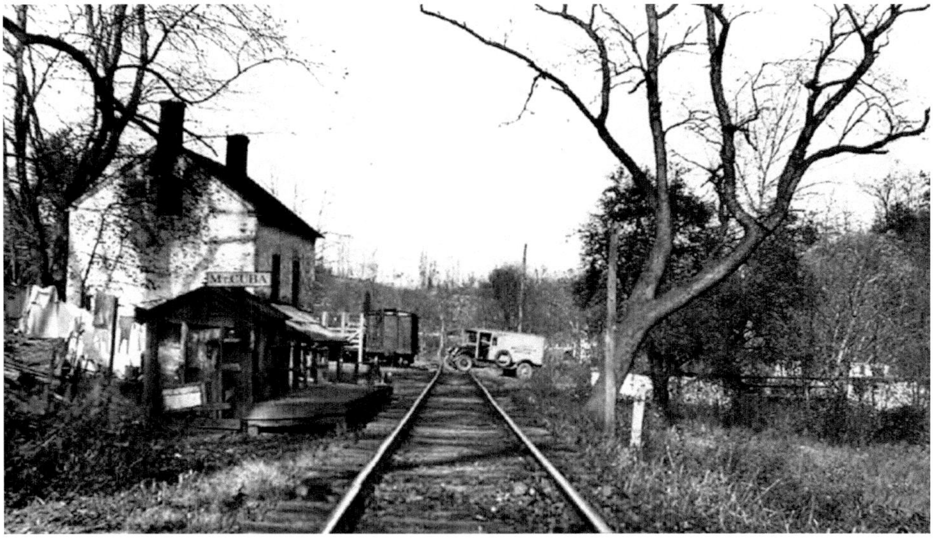

Mt Cuba, Delaware, was one of the rural stops on the line. Seven years after the trains began to run, the line once again changed hands. It was renamed the Delaware Western Railroad, which became highly profitable. In the 1880s, the line was purchased by the Baltimore & Philadelphia Railroad (B&P), a subsidiary of the Baltimore & Ohio (B&O). The line became known as the "Landenberg Branch" by the B&O and was, for a time, its most profitable branch line.

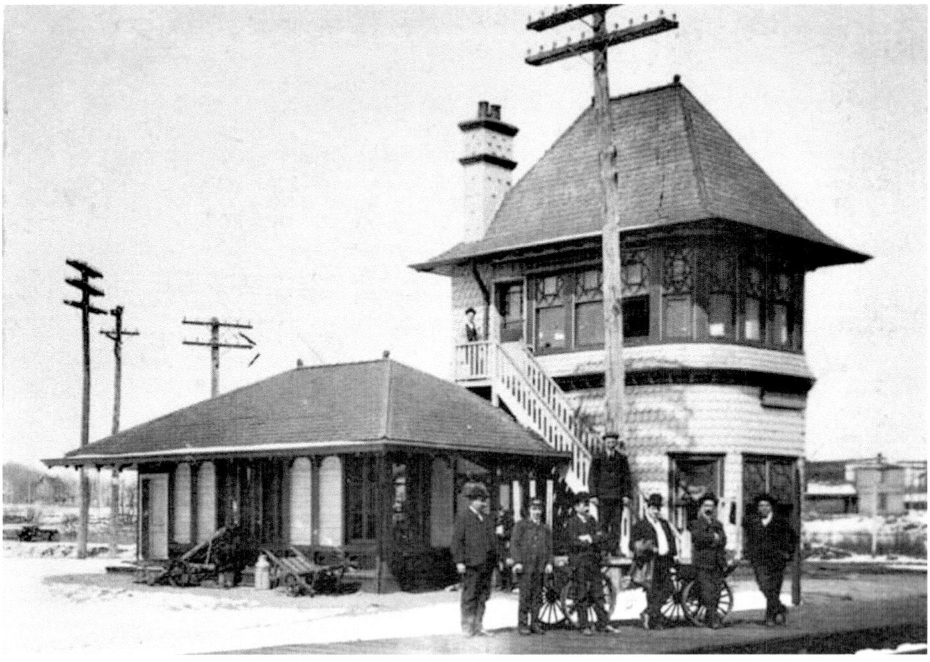

A major junction for two railroad titans was located in Elsmere, Delaware. A tower controlled the junction and diamond crossing between the Reading's Wilmington & Northern line and the Baltimore & Ohio's main line between Philadelphia and Washington, DC. With this junction and its purchase of the Wilmington and Western, the B&O finally had nonstop access to Philadelphia and cities north, a goal they had since they lost the bidding war for the Delaware Railroad with the PRR. [*Photo courtesy of Jim Bowden & Keith Short*]

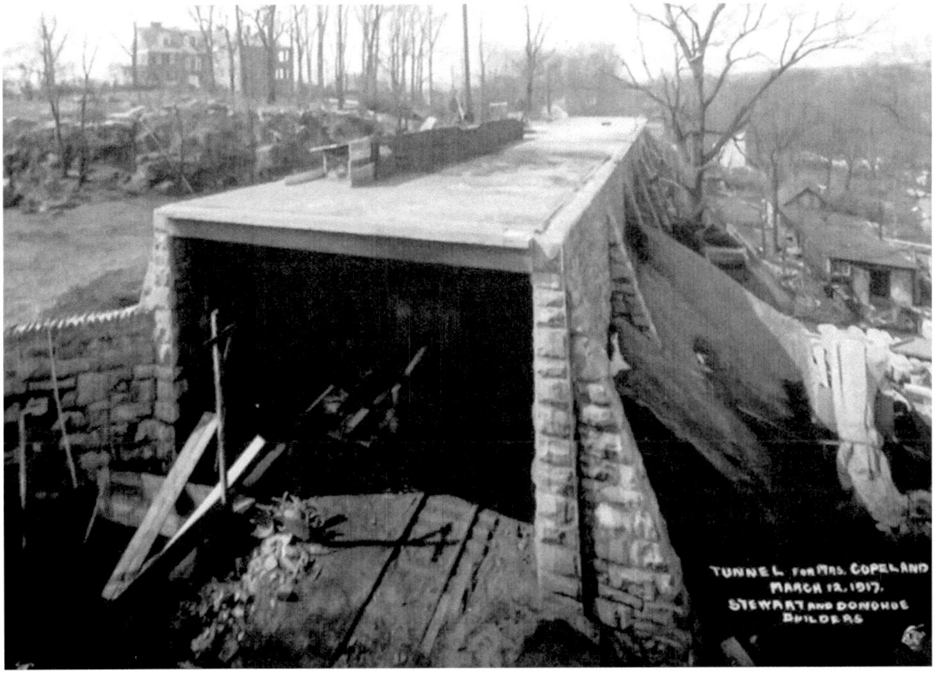

Just north of the Elsmere Junction was a small branch line that served several industries, including the "DuPont Experimental Station," a black powder plant at Hagley and textile mills such as the Joseph Bancroft Mills. The line also served various passenger stops along the way, with the last one on the line being Kentmere Station. While the line was a small portion of the rail history on Delmarva, it holds a distinction none of the others can claim as the only railroad tunnel every built in Delaware. But this tunnel was vastly different than most. It was constructed not to go under a mountain or any other natural obstacle, but the entire project was because the laundry was getting dirty. The Copeland family requested the tunnel's conctruction to keep their yard smoke free. [*Photos courtesy of Gregory D. Pawelski*]

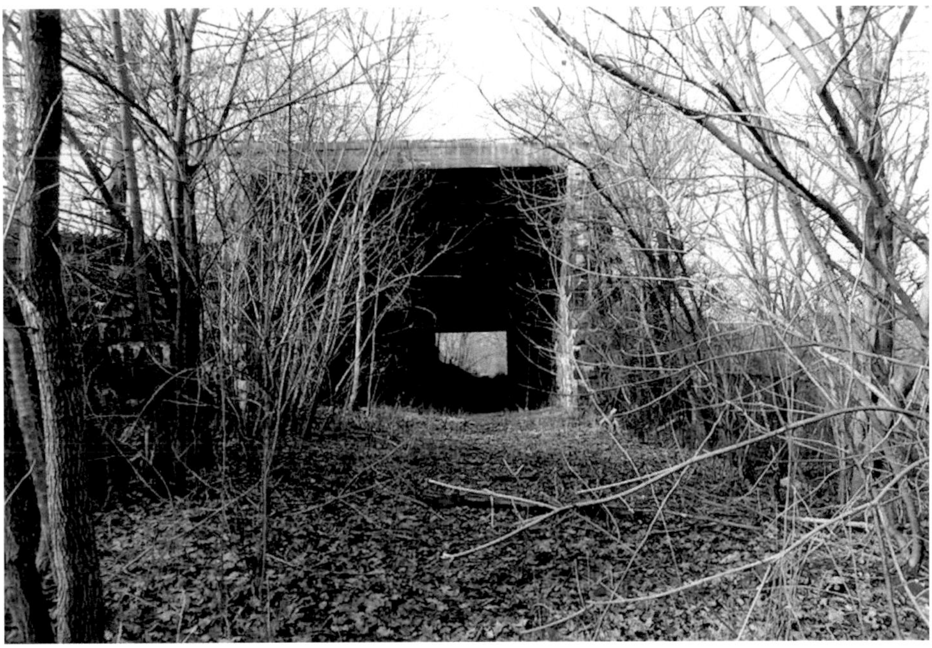

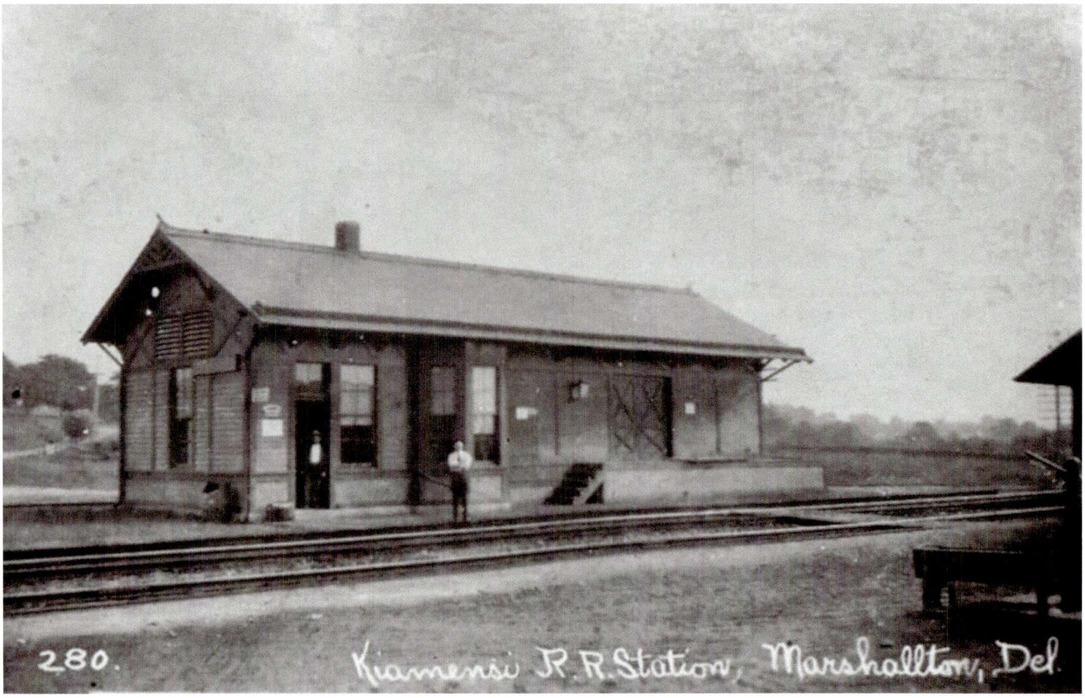

Kiamensi Station was located near Marshallton, Delaware, and the small town was home to the largest textile mill in Delaware and became an obvious choice for a railroad stop. The mill grew into a five-story structure employing over 100 workers during and post the Civil War. When the B&O took over the line, they built a new station closer to the mill. The large multistory building in the background of the new station photo is one of the many textile buildings. While all the stations are gone, the CSX still passes through the area. The Wilmington and Western serves a customer on their rail line through the connection with the CSX. [*Bottom photo courtesy of Jim Bowden & Keith Short*]

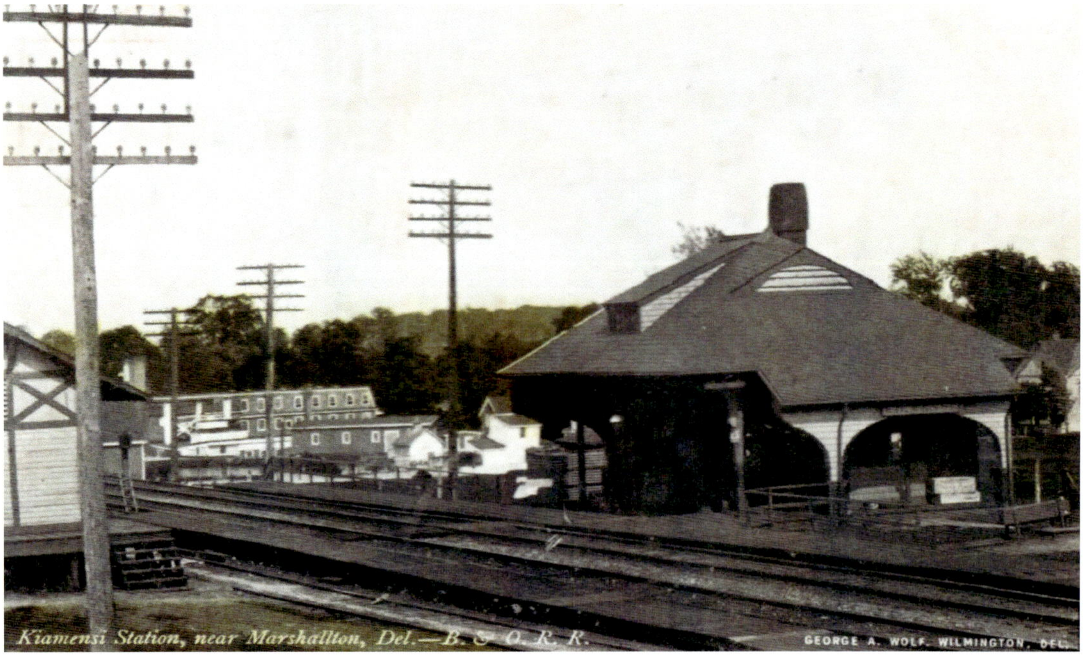

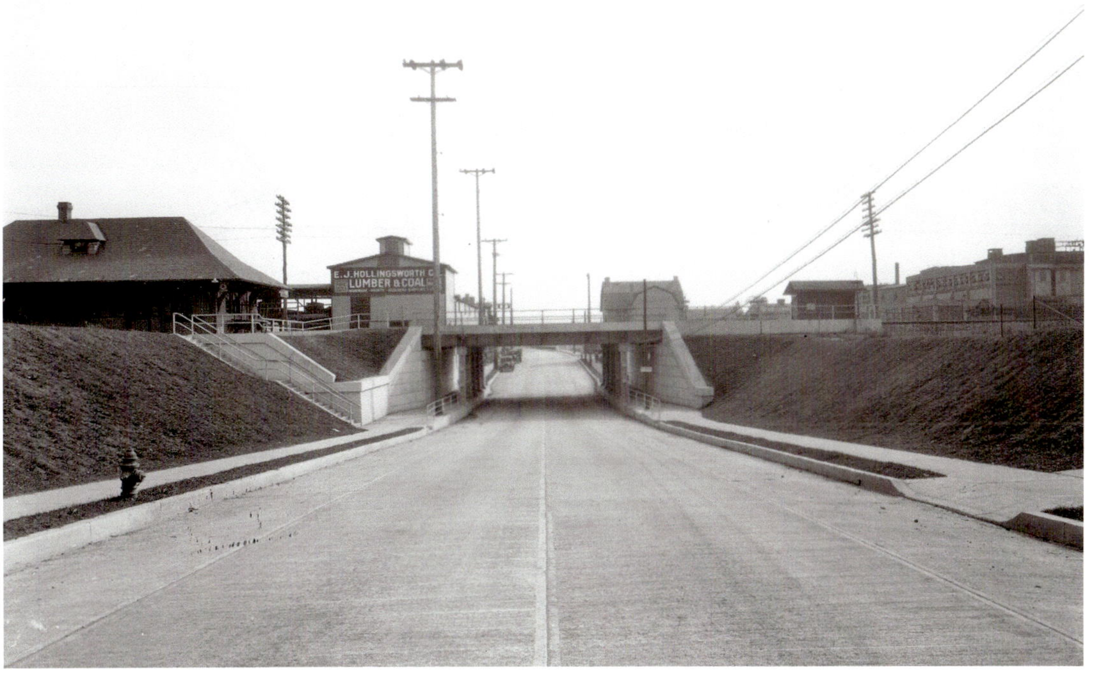

The PW&B, and later the PRR, saw the area around their stations grow in both buildings and roadway traffic at the crossing. In order to allow the trains and traffic to flow freely from each other, underpasses were a common method used to alleviate the issue. Newport, Delaware, was such a location. Originally the road crossing was at grade, but was changed to improve the traffic backups caused by the number of trains that used the rails. The station was eventually torn down, but the stairs to nowhere can be seen along the street below.

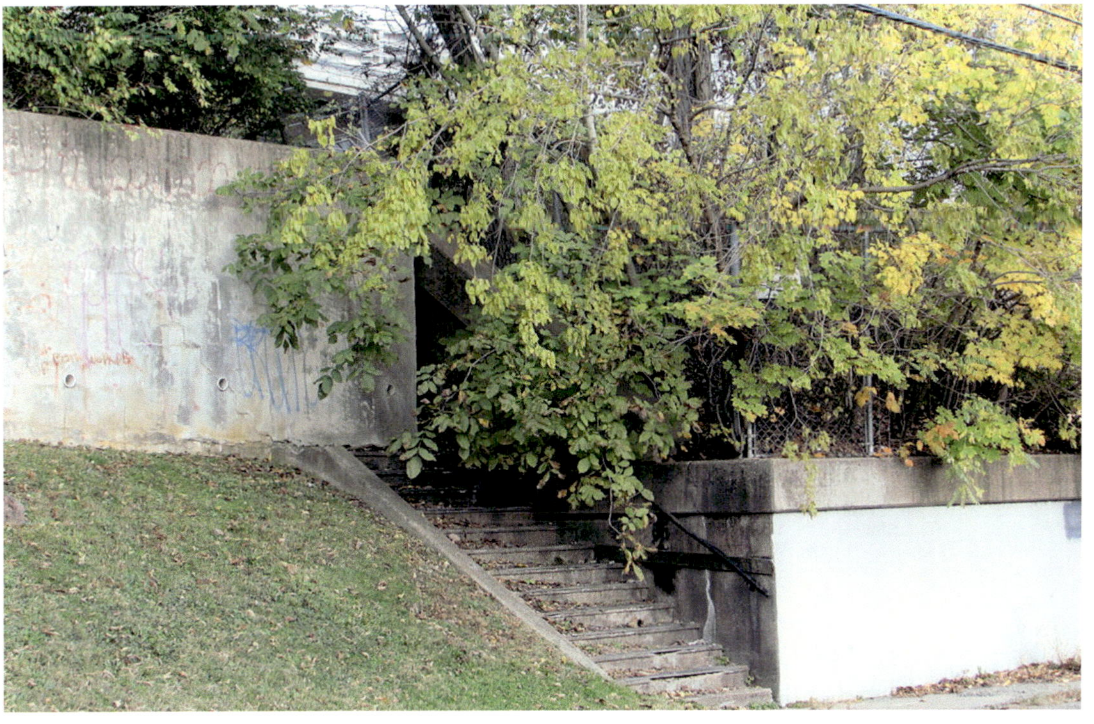

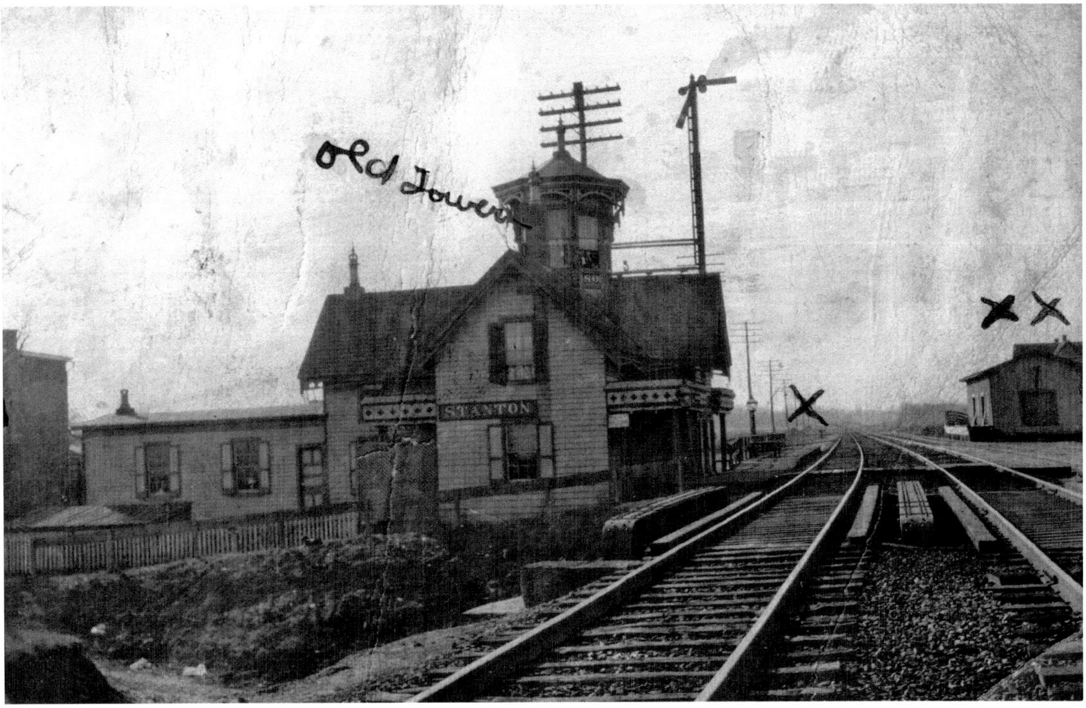

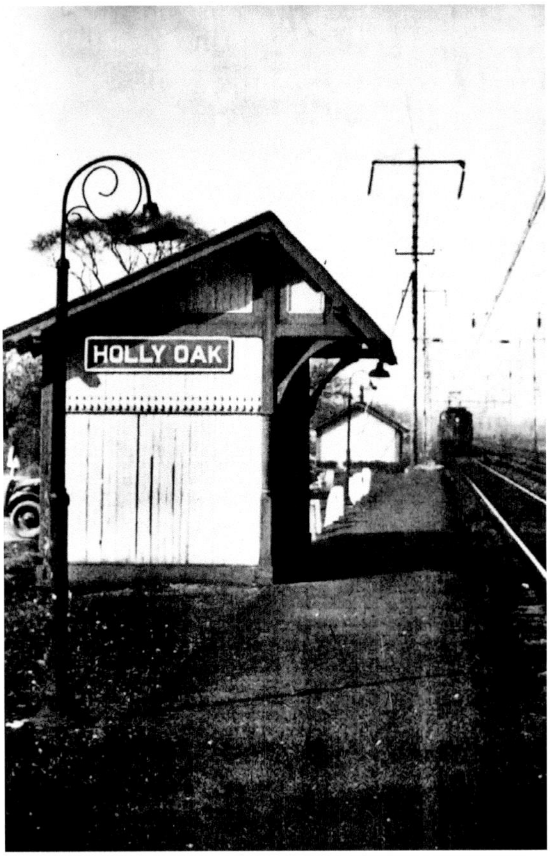

Above: Just down the line from Newport was the elegant Victorian-style station of Stanton, Delaware. Built by the PW&B, it stood through the passenger service of the PRR. The station had a unique feature of a tower on top. This gave the station master access to the semaphore signal. Today, the rails serve as a part of the Amtrak Northeast Corridor.

Left: While some stations were large and opulent, with indoor plumbing and even restaurants, some stops received what were known as post stops. Holly Oak, Delaware, was such a stop. A simple shed was constructed, open to the elements with no amenities. Even though it was small, it was important because in a time when cars were still rare, it allowed locals to walk or bike to the station to catch a train to Philadelphia. [*Photo courtesy of Robert Biddle*]

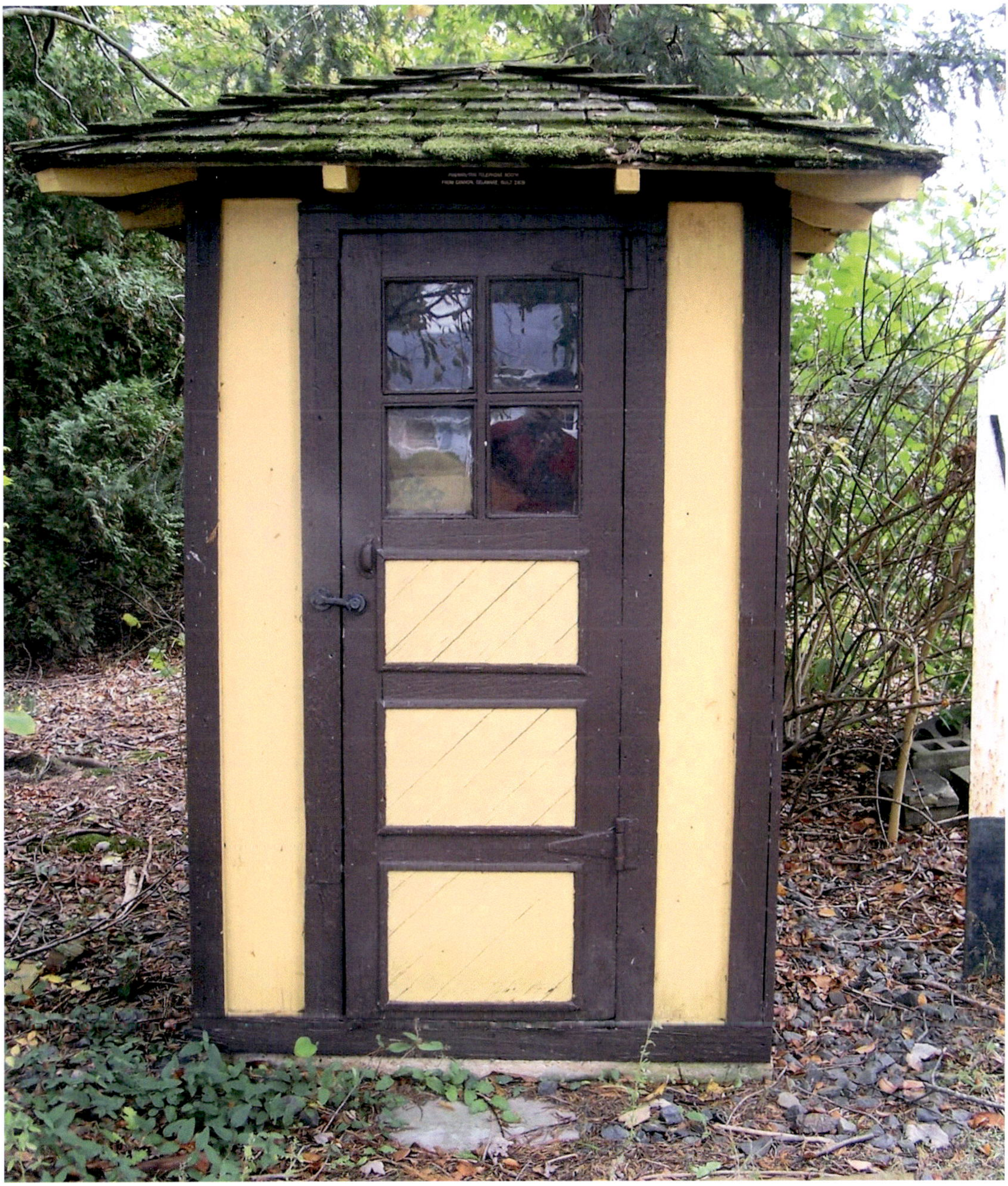

Passenger trains, freight cars, and steam engines clacking down the tracks were the romantic side of the railroads. Behind the scenes, so many more people and technologies have been used since the first railroad in the 1800s. One important need was to get orders to the trains. Using the new invention of the telephone, the PB&W installed a communication line along the railroad and placed phone booths at the stations. Surviving today, the original phone booth built in 1909 at the Cannon Delaware station reminds visitors of the time of phones before today's wireless communications. [*Photo courtesy of Jim Bowden & Keith Short*]

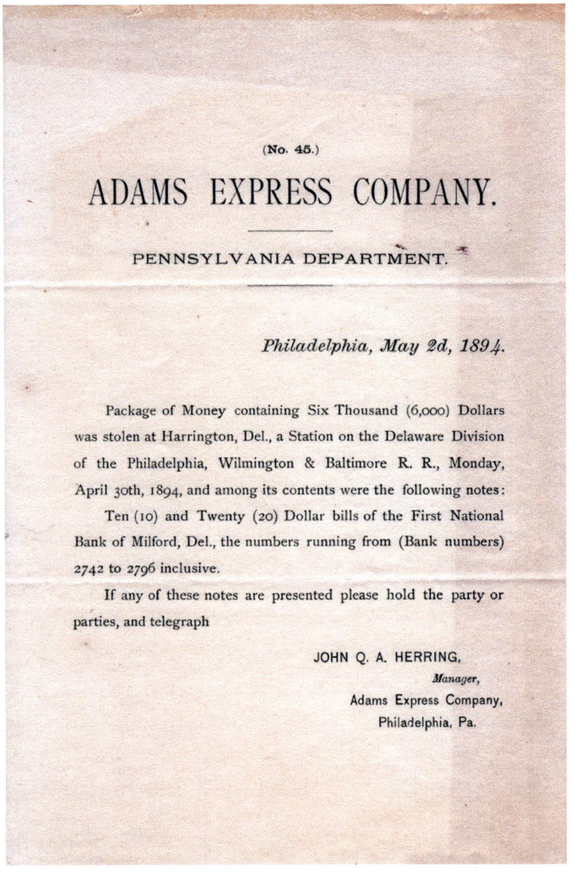

(No. 45.)

ADAMS EXPRESS COMPANY.

PENNSYLVANIA DEPARTMENT.

Philadelphia, May 2d, 1894.

Package of Money containing Six Thousand (6,000) Dollars was stolen at Harrington, Del., a Station on the Delaware Division of the Philadelphia, Wilmington & Baltimore R. R., Monday, April 30th, 1894, and among its contents were the following notes:

Ten (10) and Twenty (20) Dollar bills of the First National Bank of Milford, Del., the numbers running from (Bank numbers) 2742 to 2796 inclusive.

If any of these notes are presented please hold the party or parties, and telegraph

JOHN Q. A. HERRING,
Manager,
Adams Express Company,
Philadelphia, Pa.

Prior to UPS, Fed Ex, and even the U.S. Postal Service, there were companies such as Adams Express. Delivering everything from small packages to kits to construct your home, Adams used the ever-growing railroads to transport packages shipped by large and small companies. Cash and stocks were one of the major items carried, including payrolls for the U.S. Military. Packages arriving to the local stations would then be delivered by the local Adams agents. By the 1880s, the company had nearly 8,000 employees, was operating over 20,000 miles of railroad track, and had offices all around the United States and in most of the major cities in Europe. As the postal service began to deliver packages, the company withdrew from the package delivery service. Today, the Adams Express Company is a diversified equity investment company. [*Bottom photo courtesy of Jim Bowden & Keith Short*]

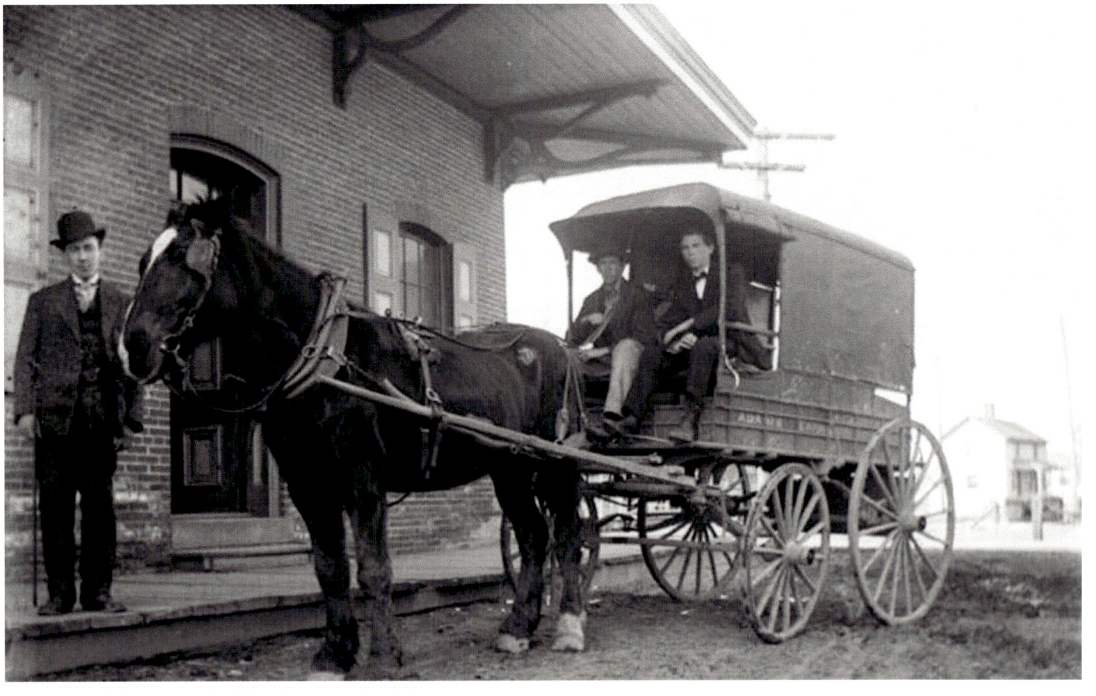

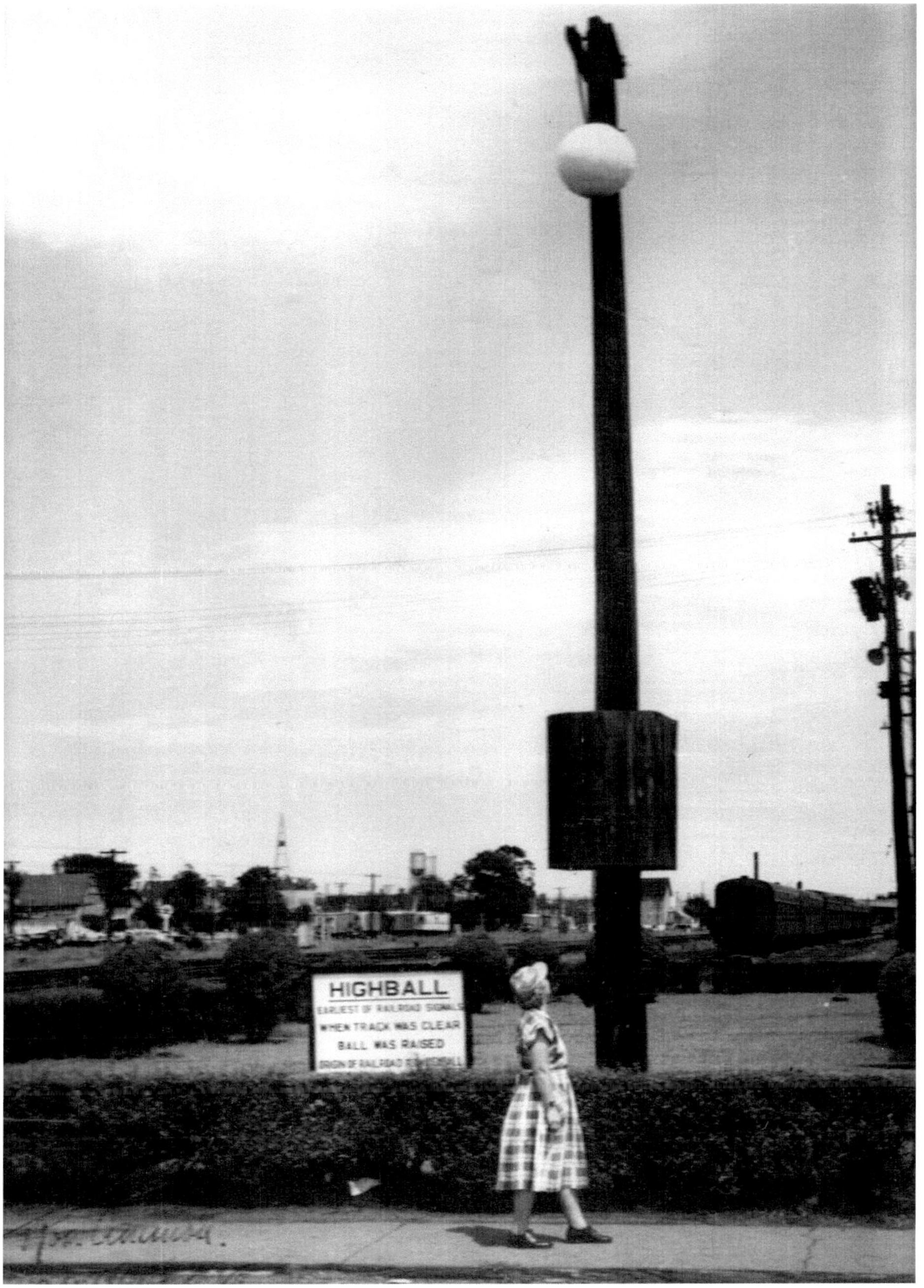

The Highball Signal was first used in Delaware on the New Castle and Frenchtown Railroad. The purpose of the signal was to give permission for a train to proceed at full speed, if the ball was at the top of the pole. The term "highball" came to be synonymous with a clear right-of-way and for trains to proceed at full speed. The one located in Delmar began its usage on the New Castle and Frenchtown, made its way to Hurlock, Maryland, and eventually found its home in Delmar, Delaware, where it was on display during the town's centennial in 1959. [*Photo courtesy of the Delaware State Archives*]

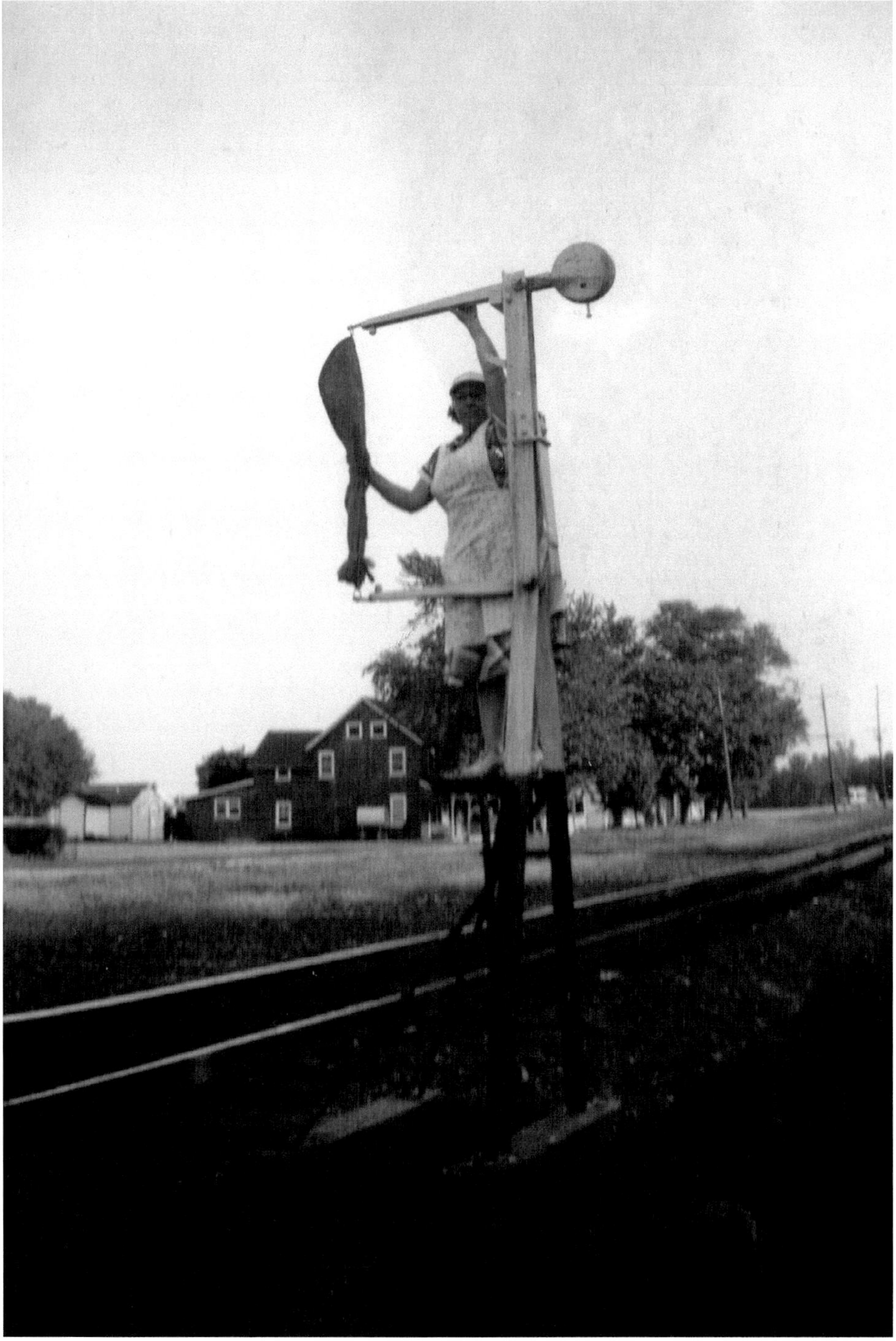

The mail seemed destined to follow the rails since almost every town clamored for the railroad in their town. Beginning in 1834, the U.S. government officially designated all railroads as postal routes. With this new service, the mail had to be picked up without stopping at each location. Enter the mail hook and catcher pouch. These allowed a catcher pouch to be hung and picked up by a passing train without the train having to stop. This became known as "mail on the fly." As the mail bag would be picked up, the worker would kick out the arriving mail.

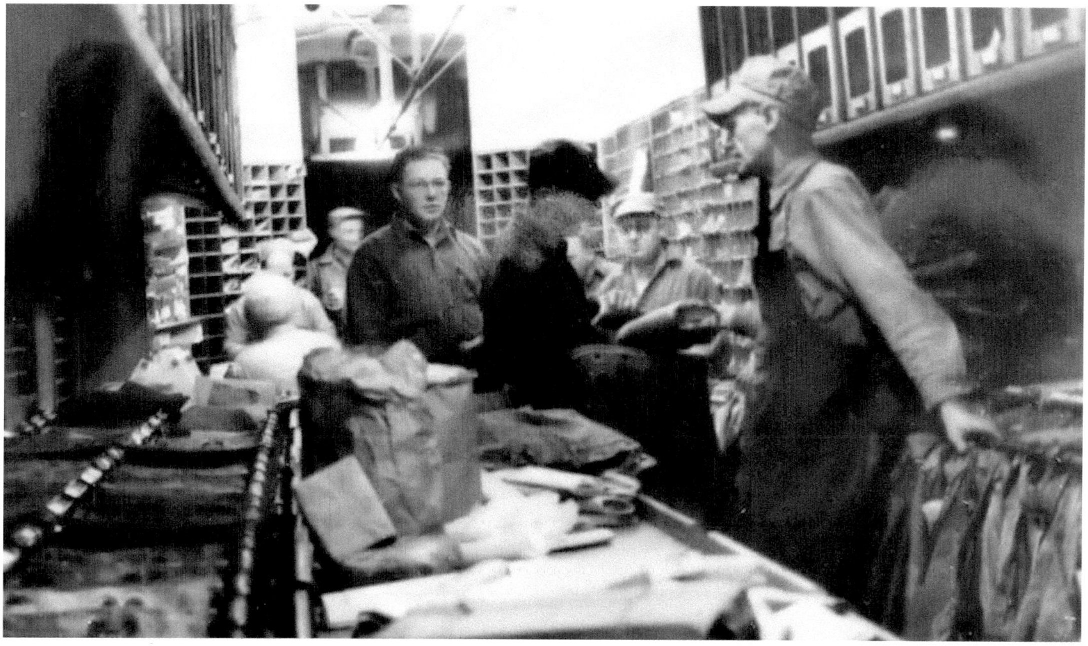

The Railway Mail Service was created in 1869 as the way to move the mail and sort the letters and parcels while on the move. The first mail cars were just a part of a converted passenger car, but soon specialist equipment and rail cars were created to improve safety and increase efficiency. RPO cars were equipped and staffed to handle postal processing. Sorting by stations and junctions, the employees had to be 96% accurate or face discipline. Initially the foreman of the crew, which could number as many as ten men, was required to carry a pistol to prevent theft of the mail. First-class mail, magazines and newspapers were all sorted, cancelled when necessary, and dispatched to post offices in towns along the route. When the federal government ended this service, many railroads, who had used the income to supplement passenger trains, saw both come to a halt.

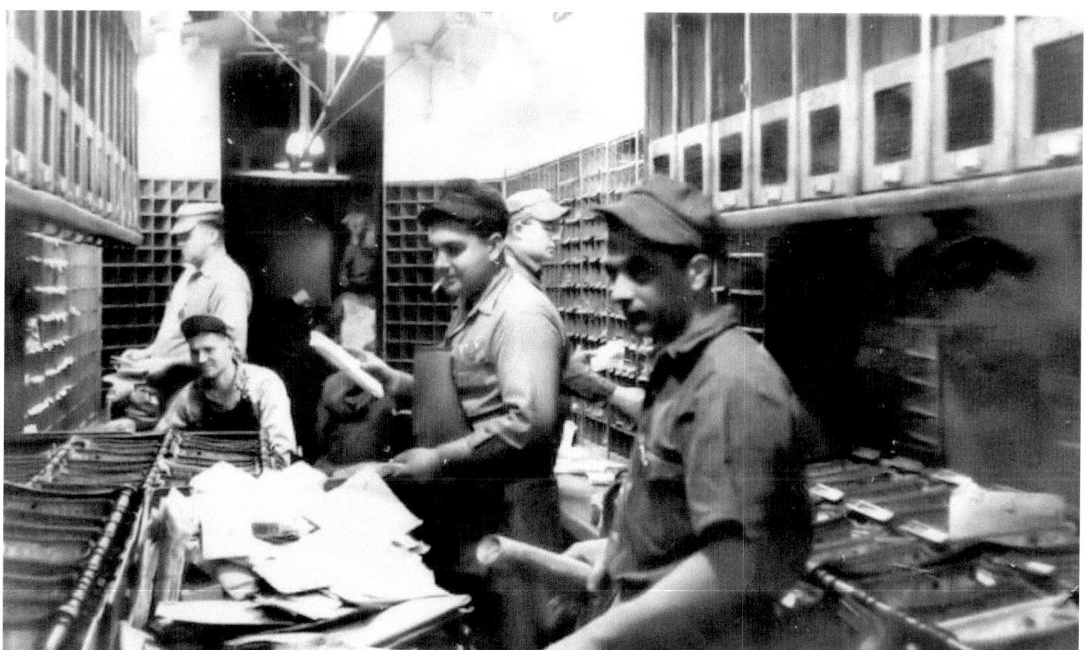

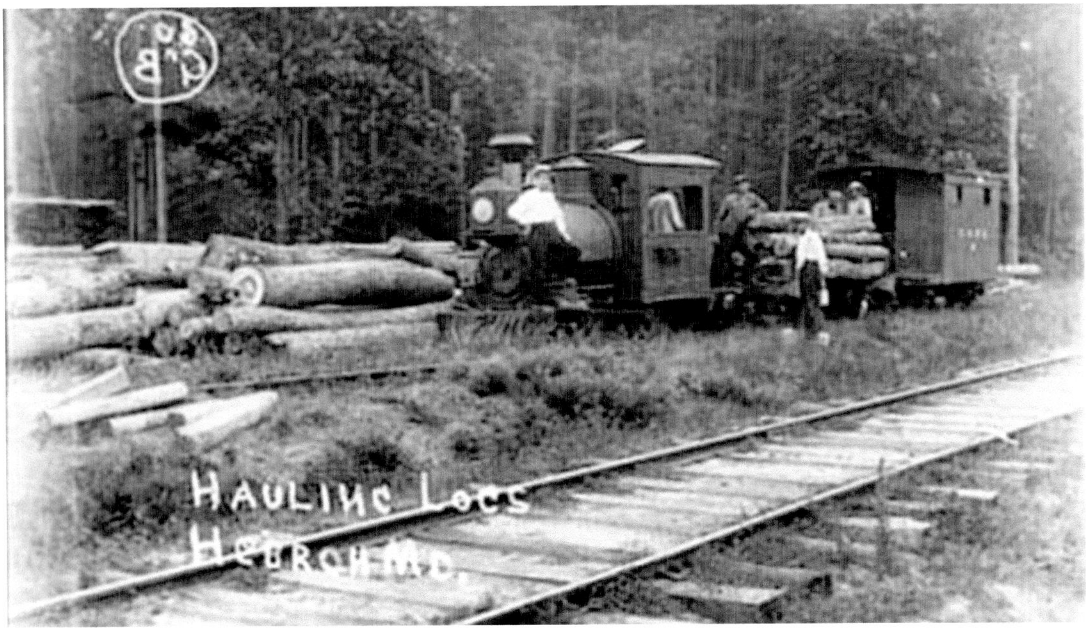

So many view the railroads as large steam trains of the past or the large diesel engines of today. That, however, does not truly tell the whole story of the rails. Narrow-gauge railroads would be constructed where normal railroads would be impossible to construct due to cost, landscape, or if the railroad was just to be used for a temporary purpose. For example, temporary railways would be removed after short-term use, such as logging, mining, or large-scale construction projects. This narrow-gauge line was built during logging operations in Hurlock, Maryland. [*Photo courtesy of William Miller*]

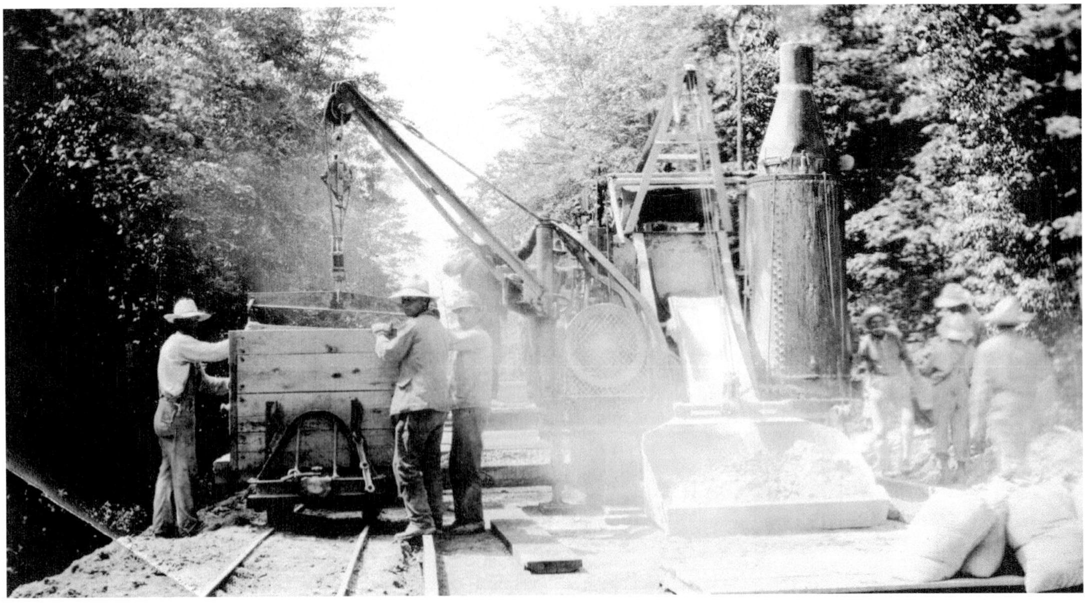

When Delaware began constructing major roadways throughout Kent and Sussex counties, a narrow-gauge railway would be constructed to bring or remove supplies and earth to and from the site such as the one at Ellendale, Delaware. Once the roadway was completed, the railway was removed. Such railways have almost vanished, however, due to the capabilities of modern trucks. [*Photo courtesy of the Delaware Public Archives*]

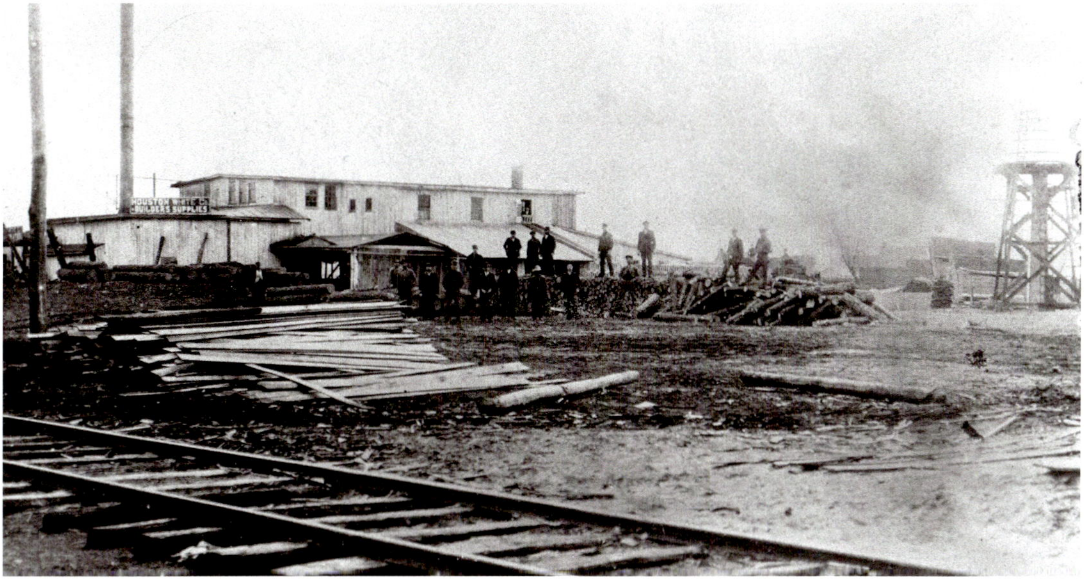

Mills and factories either already existed in areas the train crossed through or they were quickly built to allow for movement of goods and supplies. Basket factories and lumber mills, such as the Houston-White Company in Millsboro, Delaware, were some of the earliest to find their homes along the rails. Delmarva was rich in timber and baskets were needed in large quantities to harvest and ship the vast types of produce grown across the area. The small strawberry cups, the large baskets for potatoes, and even crates for things like squash, were mass produced. Soon, the canneries made an appearance. Canning was not a new process, but thanks to several innovations, almost any produce and seafood could be canned. At its peak, there were over 350 canneries and packing houses in Delmarva. Places such as Webster Canning Company in Cheriton, Virginia, processed millions of pounds of goods and shipped them across America using the railroads. Today, all these rails and factories sit abandoned.

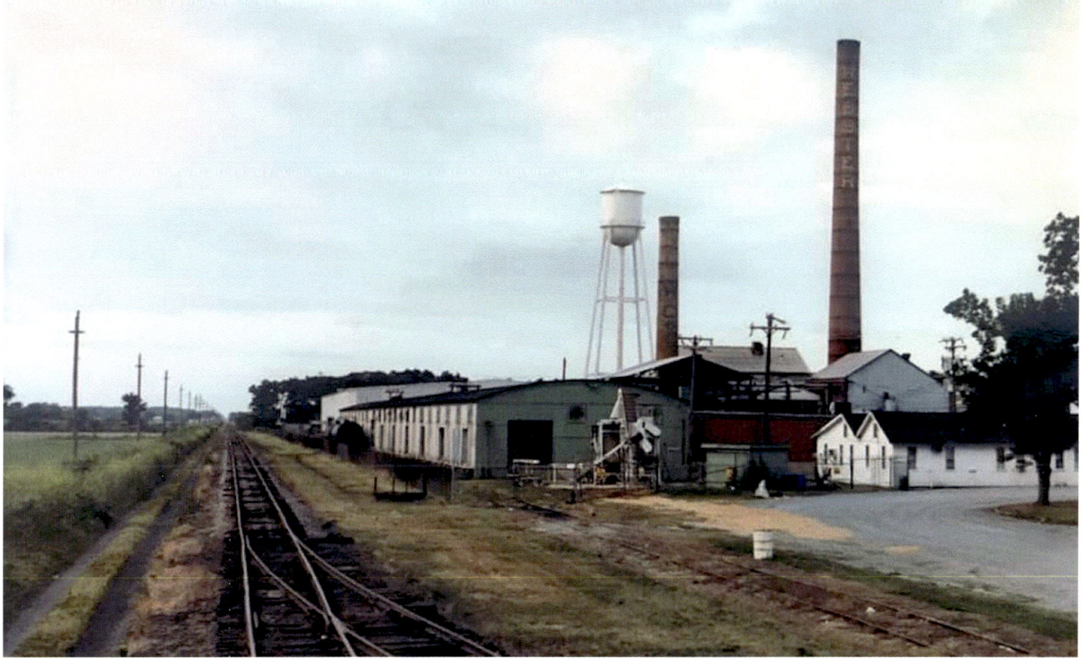

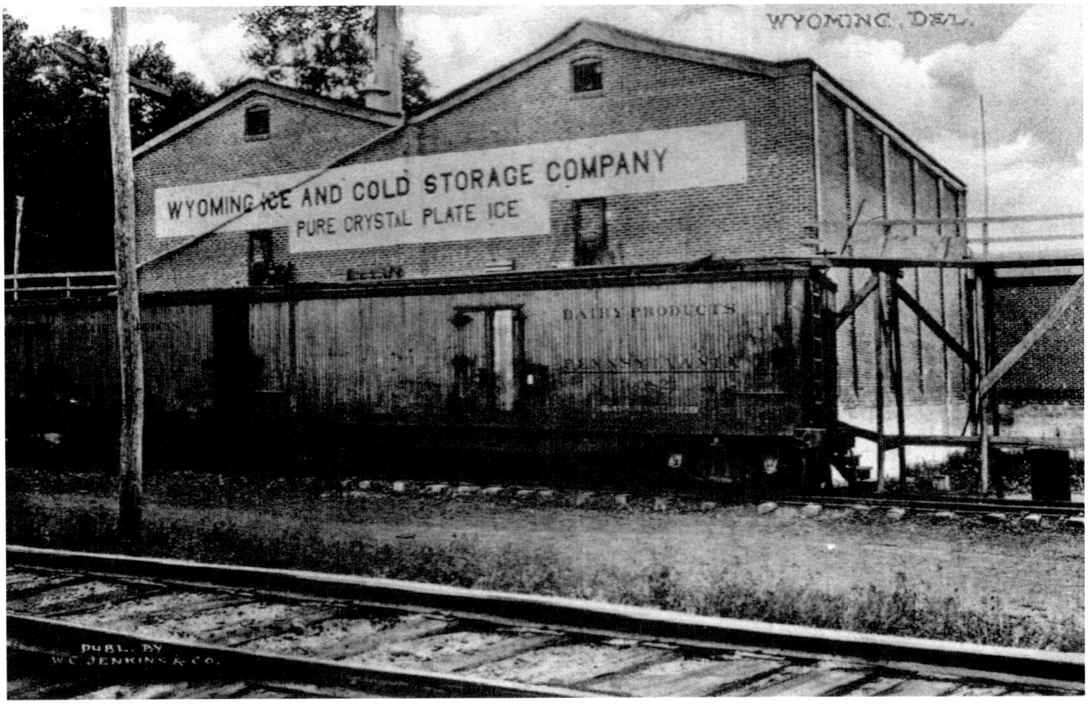

The refrigerated boxcar was truly an invention that changed America's eating habits. While some foods could be canned, almost any food could be shipped if kept cold. Originally designed for meat packers, the cars were quickly adapted for produce and seafood. And the Delmarva Peninsula was perfectly situated to take advantage of the new shipping method. Now, a person in New York could taste fresh sweet corn, a strawberry that had been picked the day before, and crabs still alive that had to be steamed and eaten, giving them tastes they had never known. The favorites of big cities were rockfish and fresh oysters, shipped by the ton. Ice houses like the one in Wyoming, Delaware, were built directly next to the rails and included a second story with a walkway. Men would push two- to four-inch blocks of ice into the hatches on the roof of the cars, mix it with salt, and the goods in the cars would be refrigerated for a day. Dairy products would also become a commodity shipped to the cities, as milk and other dairy products could now be kept cold and fresh. These ice houses soon sprang up in cities up and down the Delmarva Peninsula. Today, most of these are gone or have been turned into warehouses.

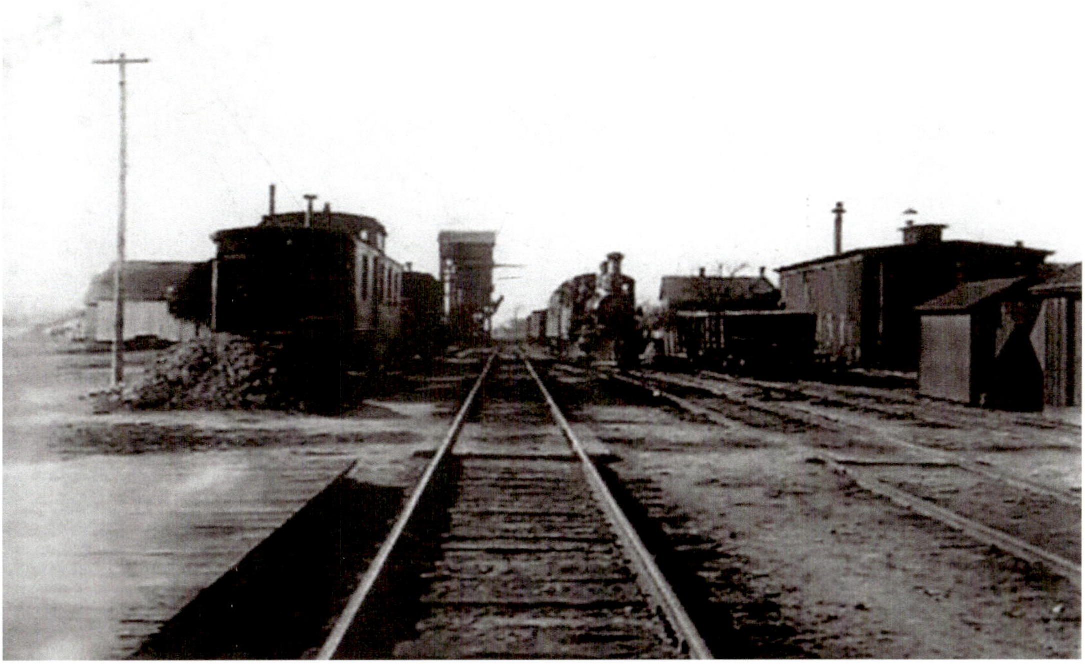

Maintenance on the railroad was a twenty-four-hour-a-day job. Places such as Georgetown, Delaware, had one- or two-stall engine houses for repairs, while larger years such as those in Wilmington or Cape Charles had multi-stall houses with large roundtables so work could be performed on multiple engines at one time. During the steam era, boiler work, lubrication, and inspections were daily occurrences. Diesel locomotives reduced those requirements, especially the lubrication of the large drive wheels that were essential to steam locomotives. Most engine houses have been removed as the maintenance has been centralized to a few locations; however, some, such as the one in Georgetown, have been given a second life. This one has been turned into an apartment building.

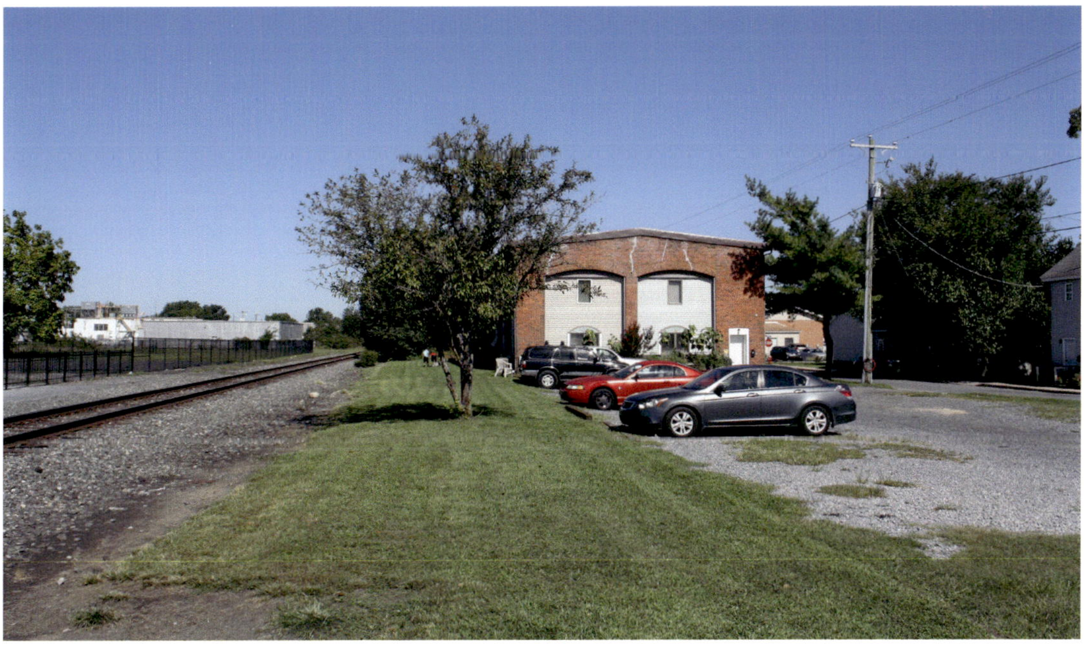

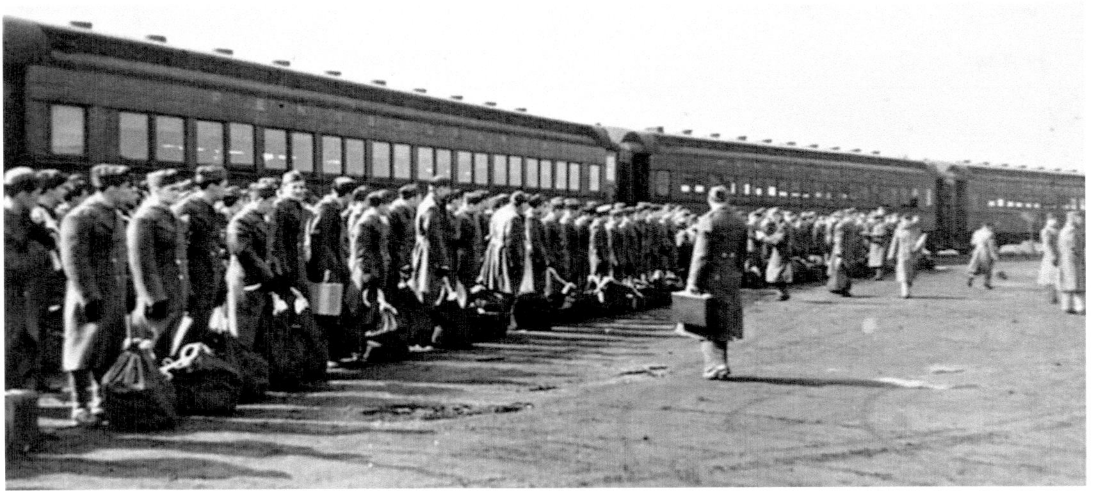

The railroads proved their might perhaps at no greater time than in wartime. With the ability to move men and supplies almost directly to the battlefield, the train became an invaluable asset to America during any conflict. During World War II, the rail lines on Delmarva moved goods and men to keep the war effort going. Scenes such as the arrival of fresh troops to Delaware City, Delaware, where they would march down to Fort DuPont at the edge of the city, were common sights. [*Photo courtesy of Robert Biddle*]

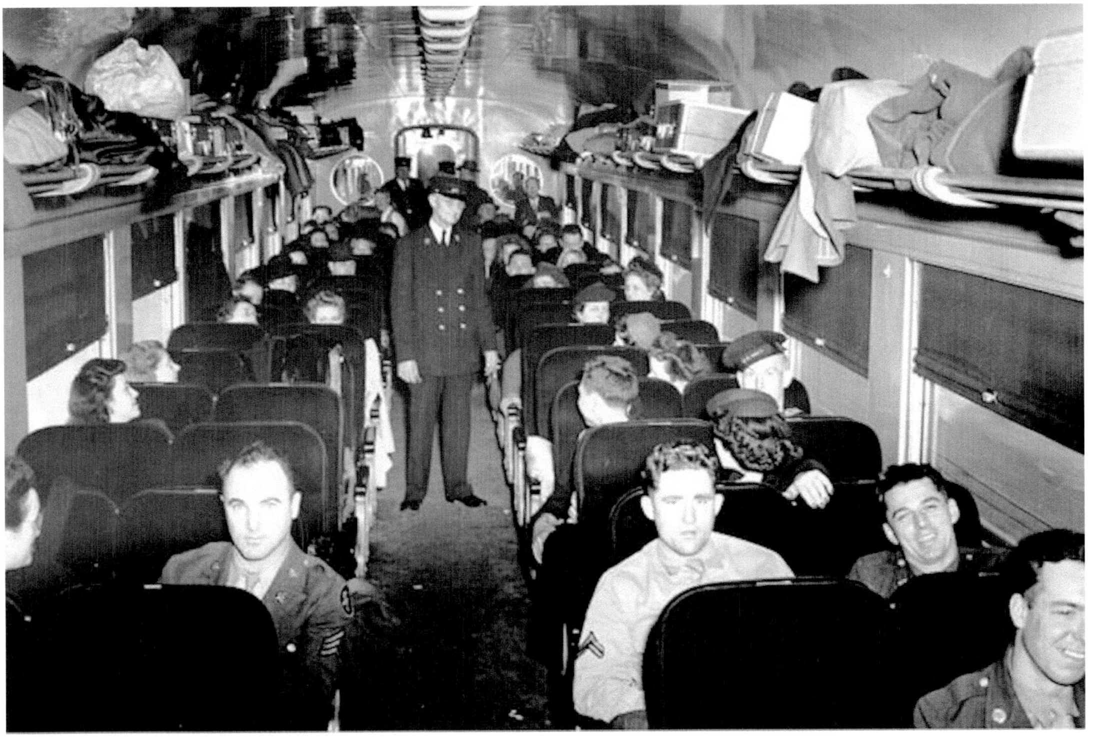

With connections to such places at the major naval bases of Norfolk, there was no shortage of military men and women traveling by train across Delmarva. At seventy years old, D. J. Truitt, a conductor for the PRR Norfolk Express, takes tickets from a train filled with servicemen and women. The ability to move these men and women quickly made the railroads on Delmarva a major contributor to our America's war efforts. [*Photo courtesy of the Delaware Public Archives*]

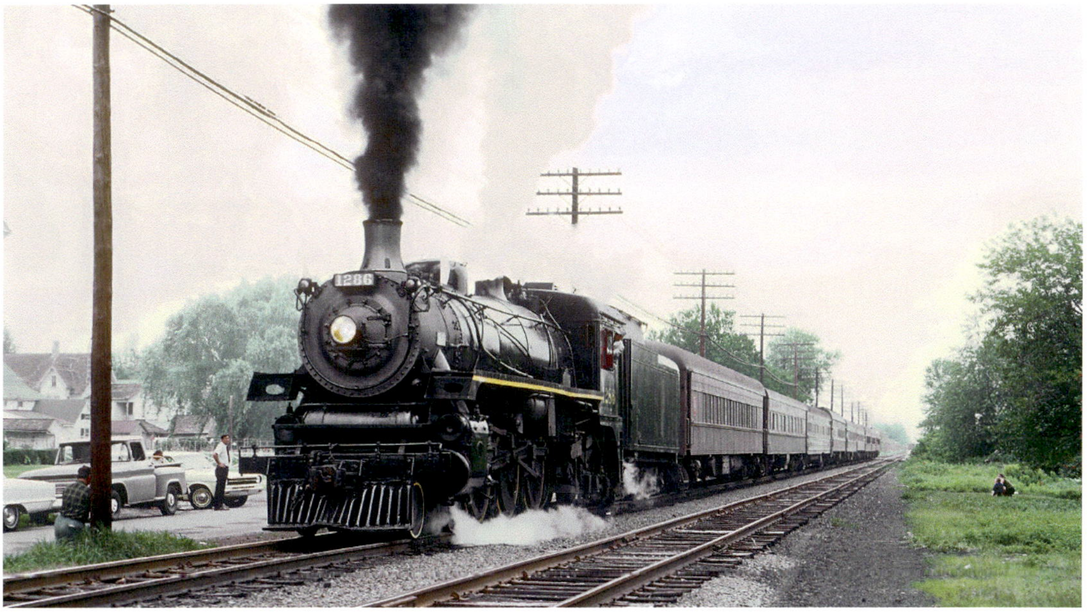

While not all rails on Delmarva were abandoned, certainly the methods used to travel the rails changed over the decades. But occasionally, one could see history come to life. One such excursion train was chartered for the Delmarva Chicken Festival, held annually from 1948 to 2014. The train would begin in Wilmington and travel the old Delaware Railroad line, ending in the community that was hosting the event. In June of 1969, the train passed through Harrington, Delaware, on its way to Seaford, Delaware. [*Photo courtesy of Jim Bowden & Keith Short*]

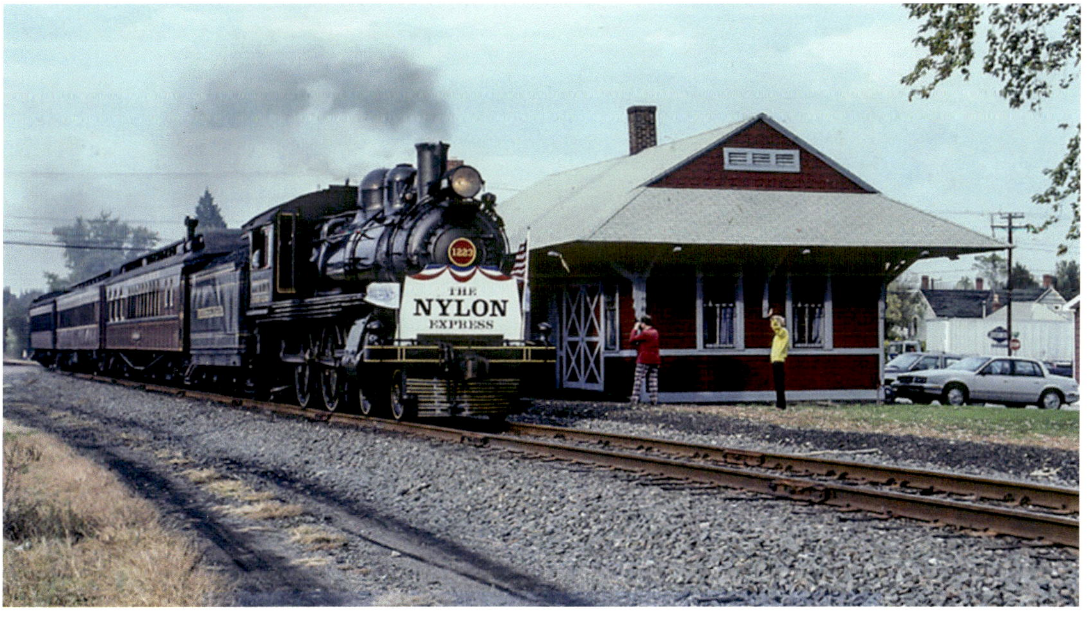

Perhaps no other company that called Delaware home is more connected to the state than the DuPont Corporation. Founded in 1802 by Éleuthère Irénée du Pont, the company was known around the world as the supplier of some of the best black powder ever made. The company would go on to such inventions as nylon, Kevlar, Teflon, and many other products used today. In 1988, a special train called the "Nylon Express" was charted to celebrate the company's fifty years of plant operations in Seaford, Delaware. [*Photo courtesy of Jim Bowden & Keith Short*]

P. R. R. Films 100th Anniversary PICTURE at Hurlock, Md.

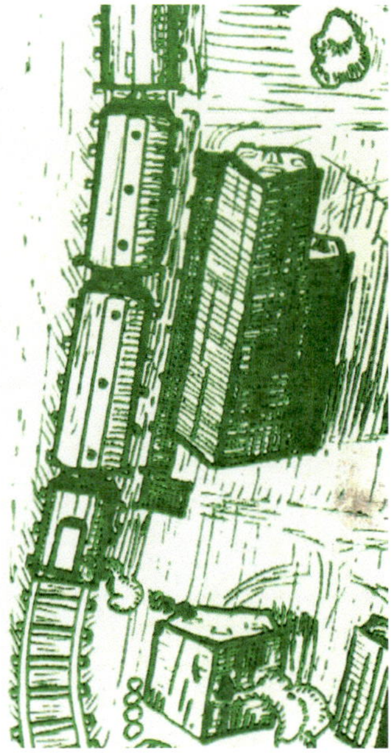

The history of the railroad was always most important to those that worked the lines. With so many older lines on Delmarva, and at the time so many stations still standing, Hurlock, Maryland, was chosen as the site to film a portion of the 100th anniversary movie about the Pennsylvania Railroad. Made in 1946, Hurlock was chosen to represent old-time railroading. Called *Clear Track Ahead*, Hurlock represented Lewiston, Pennsylvania. The thirty-minute movie highlights the changes to the railroading industry in just 100 years. A replica John Bull engine was even used to give the movie a great authenticity. The movie can be seen today by searching on YouTube.

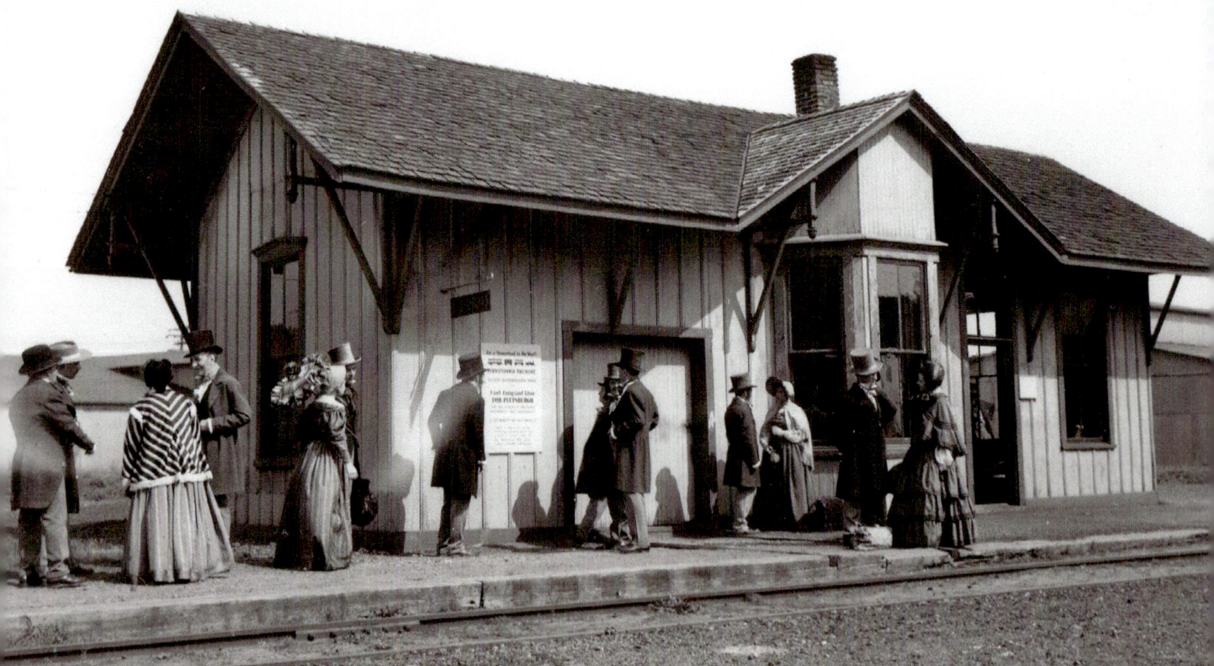

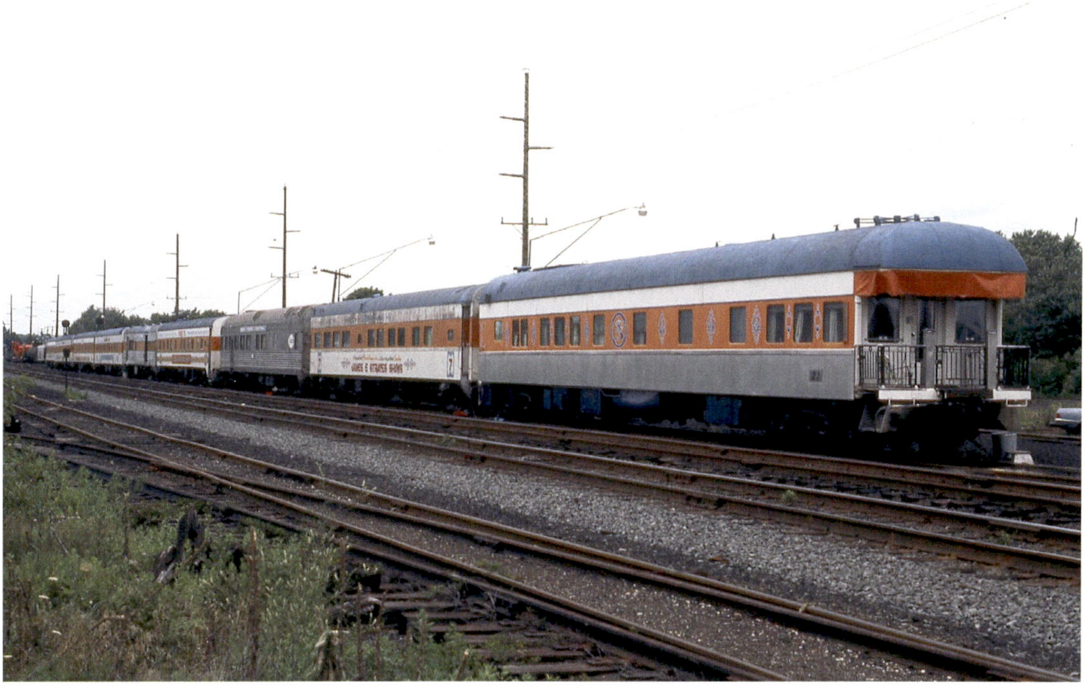

While commercial passenger service ended outside of Amtrak and Septa on the Delmarva Peninsula in 1965, one train continued to show up each summer that harkened by to those days. The Delaware State Fair has been held annually in Harrington, Delaware, each July since 1920. From the 1970s to the late 1980s, James E Strates Shows, a carnival amusement company based in Orlando, Florida, traveled by train to the railroad near the fairgrounds. Today, James E Strates is America's only amusement company that travels by rail. Locals looked each day, including this author, to see if the train had arrived, knowing it signaled the coming of the fair. Workers lived and traveled from city to city on the train. [*Photos courtesy of Rich Taylor*]

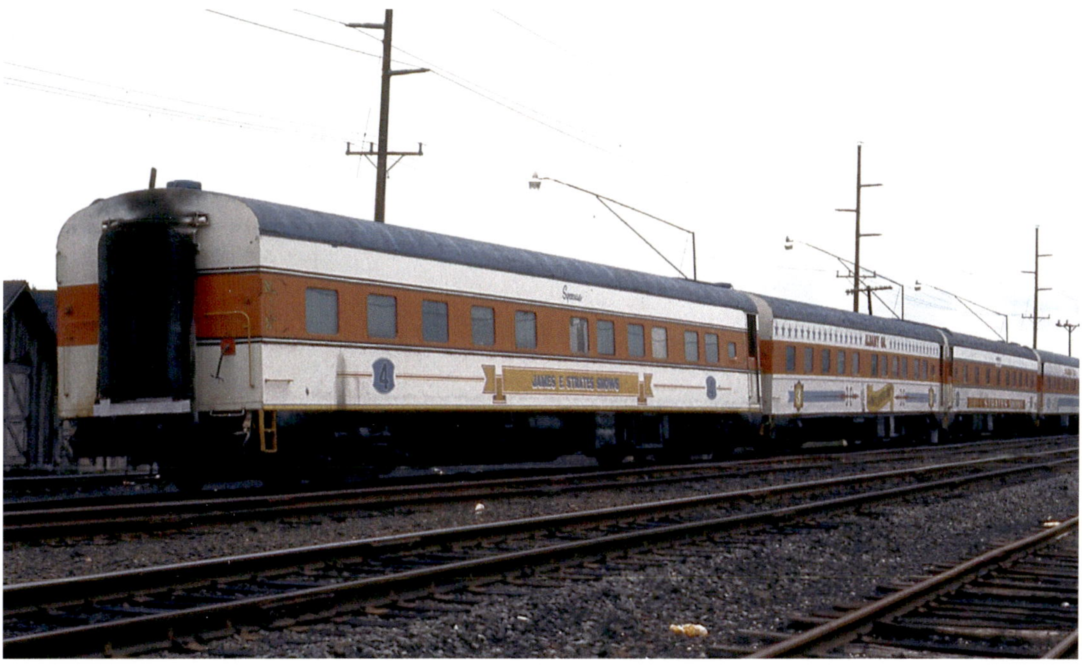

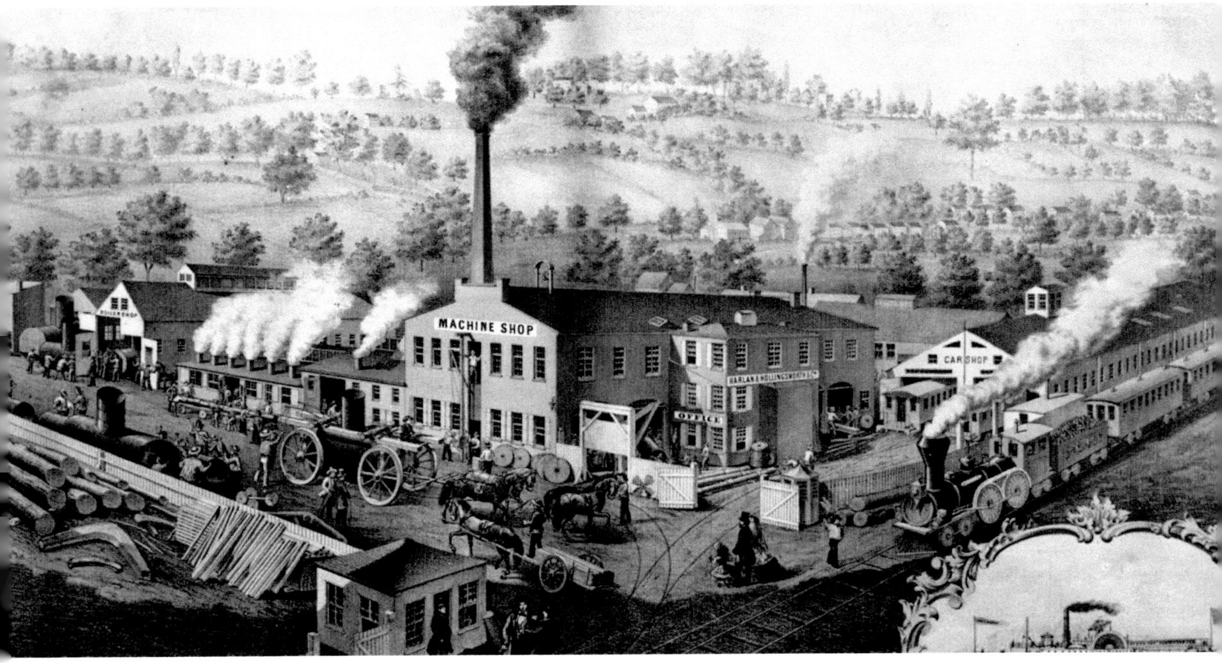

It has not just been railroads that have vanished from Delmarva, but construction of locomotives, passenger cars, and even wheels for the railroad all at one time or another were constructed in Delaware particularly. One of the largest was located in Wilmington, Delaware: Harland and Hollingsworth. The company started producing railway passenger cars in 1836 and eventually expanded to iron ship building as well in 1843. During the mid-to-late 1800s, Harlan and Hollingsworth garnered a stellar reputation for high-quality railway passenger cars, iron-hulled steamers, steam engines, and boilers.

On the next page: By 1889 the Harlan and Hollingsworth Company employed 1,500 workers, paid $1,000,000 in wages, and had fifty buildings occupying 43 acres. One of their employees would go on to national fame as a professional baseball player and then be implicated in one of baseball's most infamous scandals. In 1918, Shoeless Joe Jackson took a job in a shipyard to contribute to the war effort and played for the industrial league team. Second only to the Pullman Company in length of time building railroad passenger cars, the company was sold to the Bethlehem Shipbuilding Corporation in 1904. The company quit producing ships in 1926 and railroad car in 1939. The office building remains today and is on the National Registry of Historic Places.

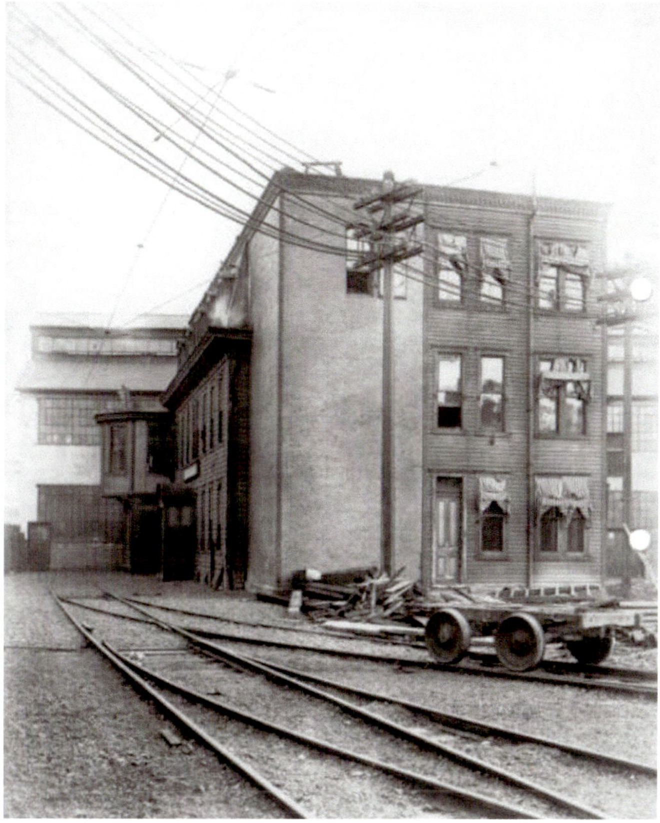

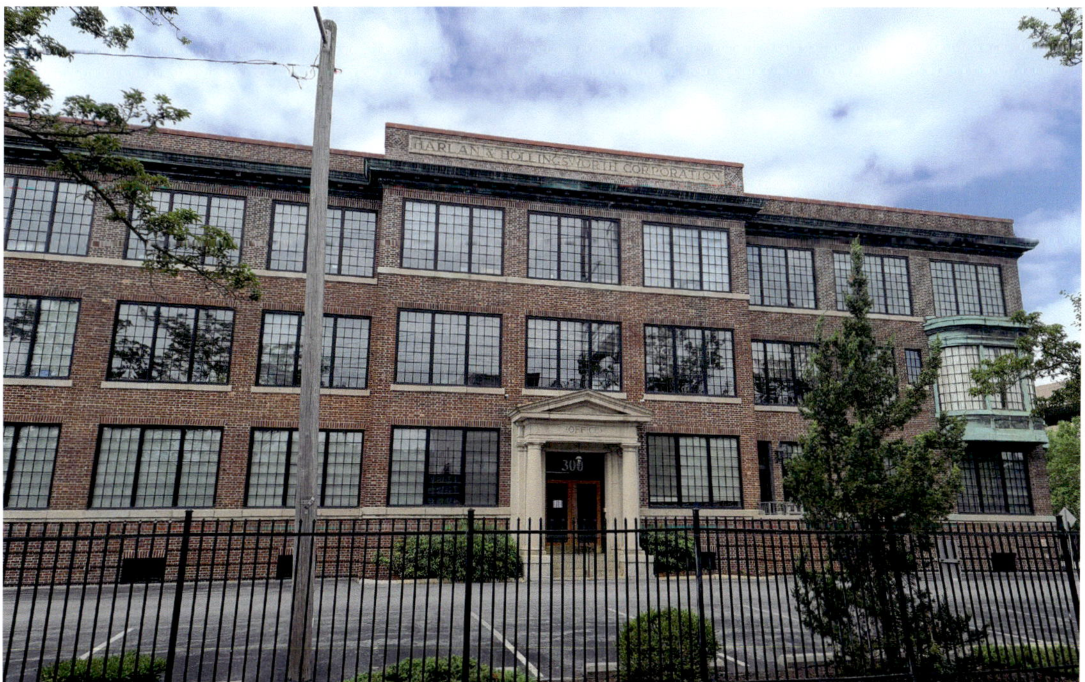
HARLAN & HOLLINGSWORTH CORPORATION

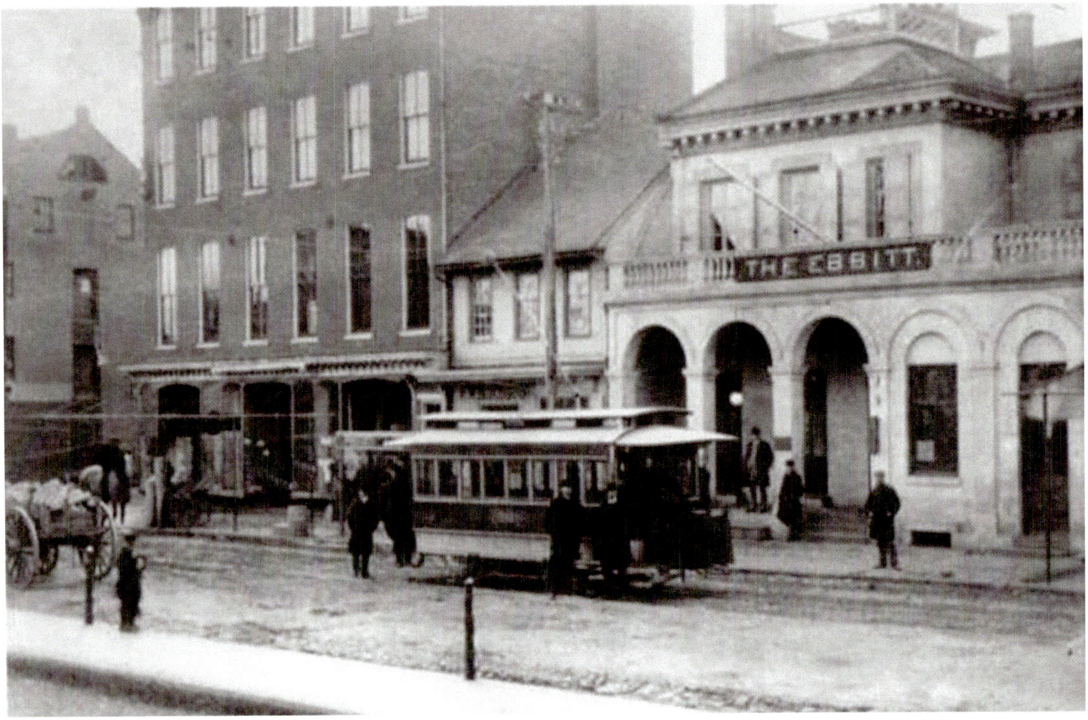

Not all the rails on Delmarva ran between cities and states. Some ran between city blocks: the trolley system. While only ever constructed in New Castle County, Delaware, over twenty trolley lines had been proposed from various spots on Delmarva. Here in front of the famous Ebbitt House in Wilmington, a trolley from the Wilmington Street Railway carries passenger around the business district of the city. The corner of 10th and Market Streets today shows a more modern city, but in the foreground, the old trolley tracks are still visible.

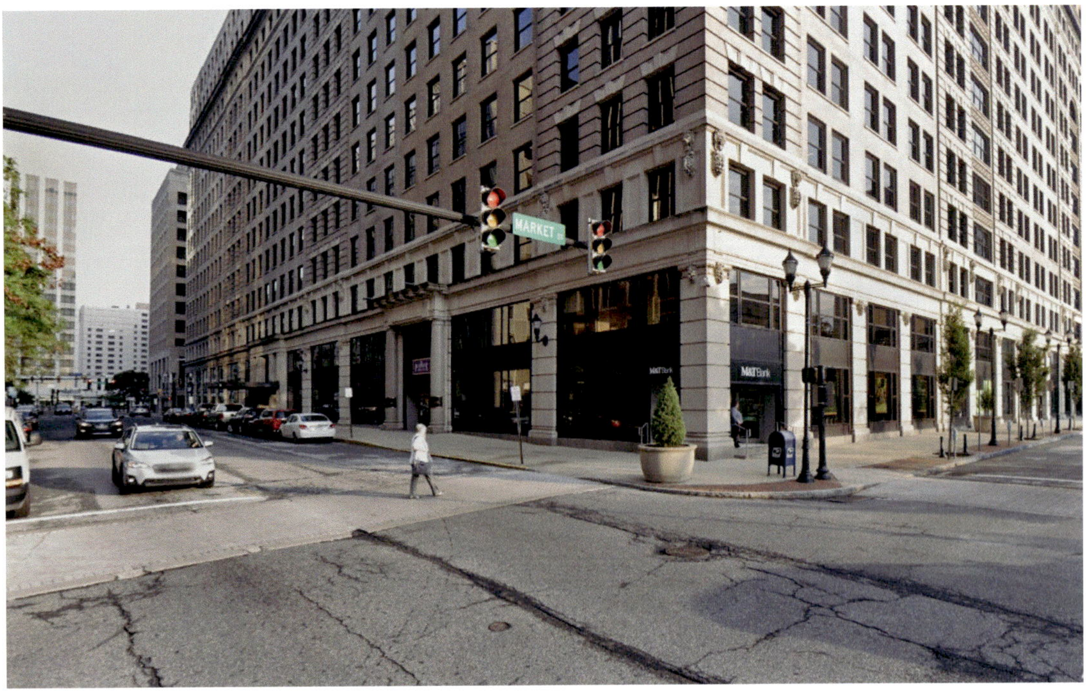

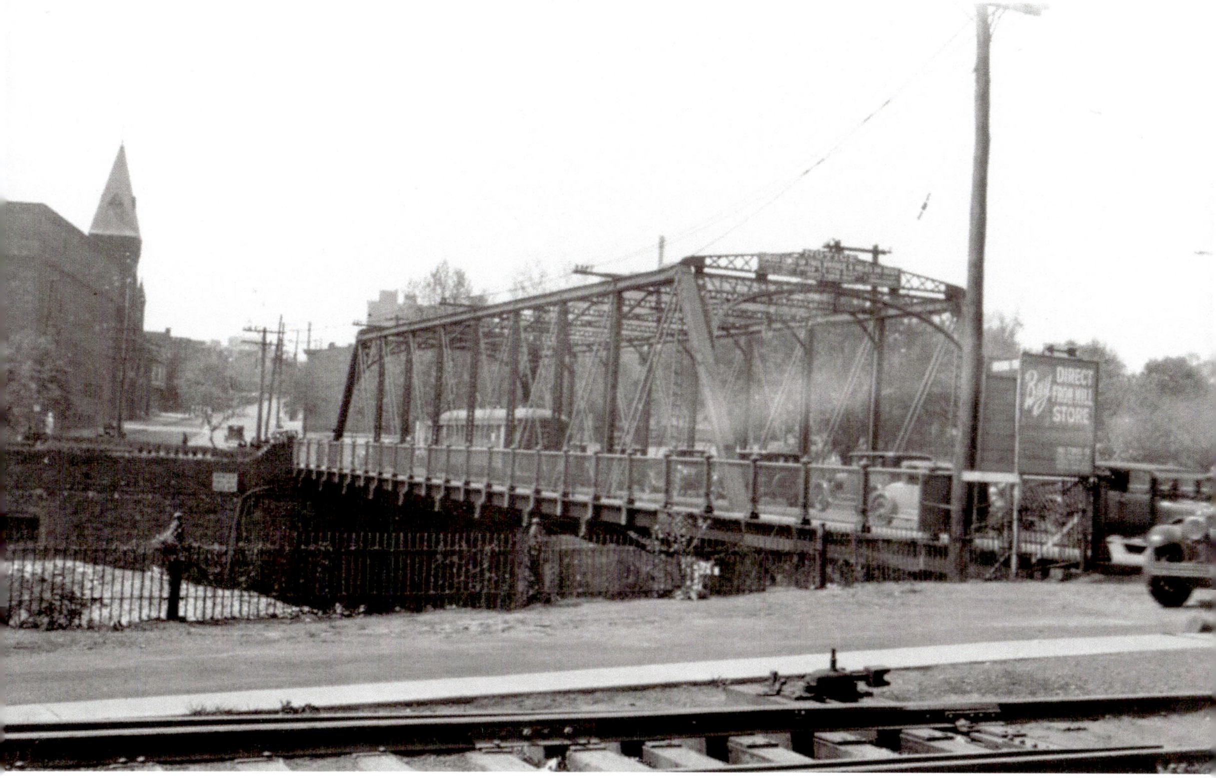

The trolley line and vehicle traffic shared the South Market Street Bridge by 1928 when this photo was taken. Delaware Electric Power Company, by this time, owned both the trolley and bus lines in the city, while also now being the company that was generating the power for homes and businesses. Eventually, the company became what is known today as Delmarva Power and supplies most of the state with its electrical needs. The bridge over the Christina River was eventually replaced and all remnants of the trolley line removed.

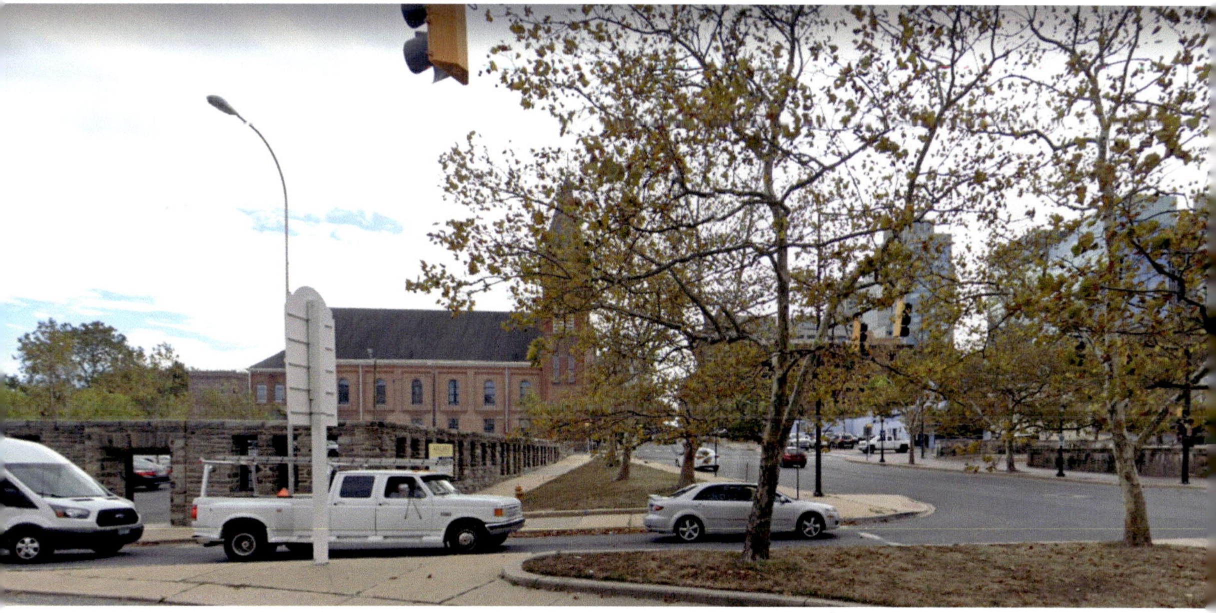

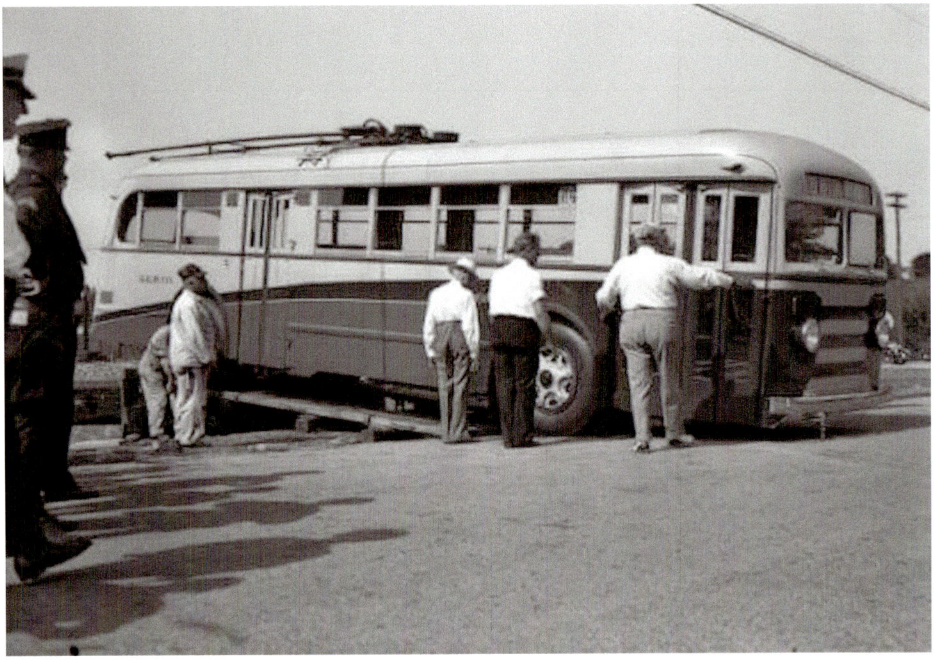

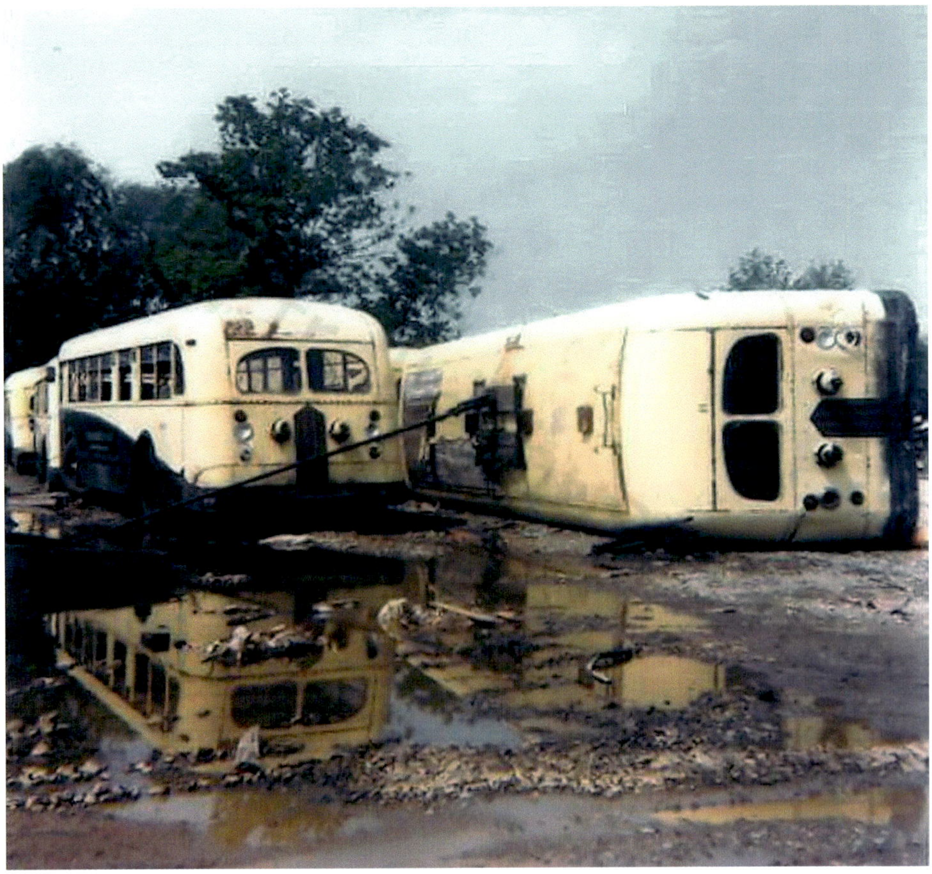

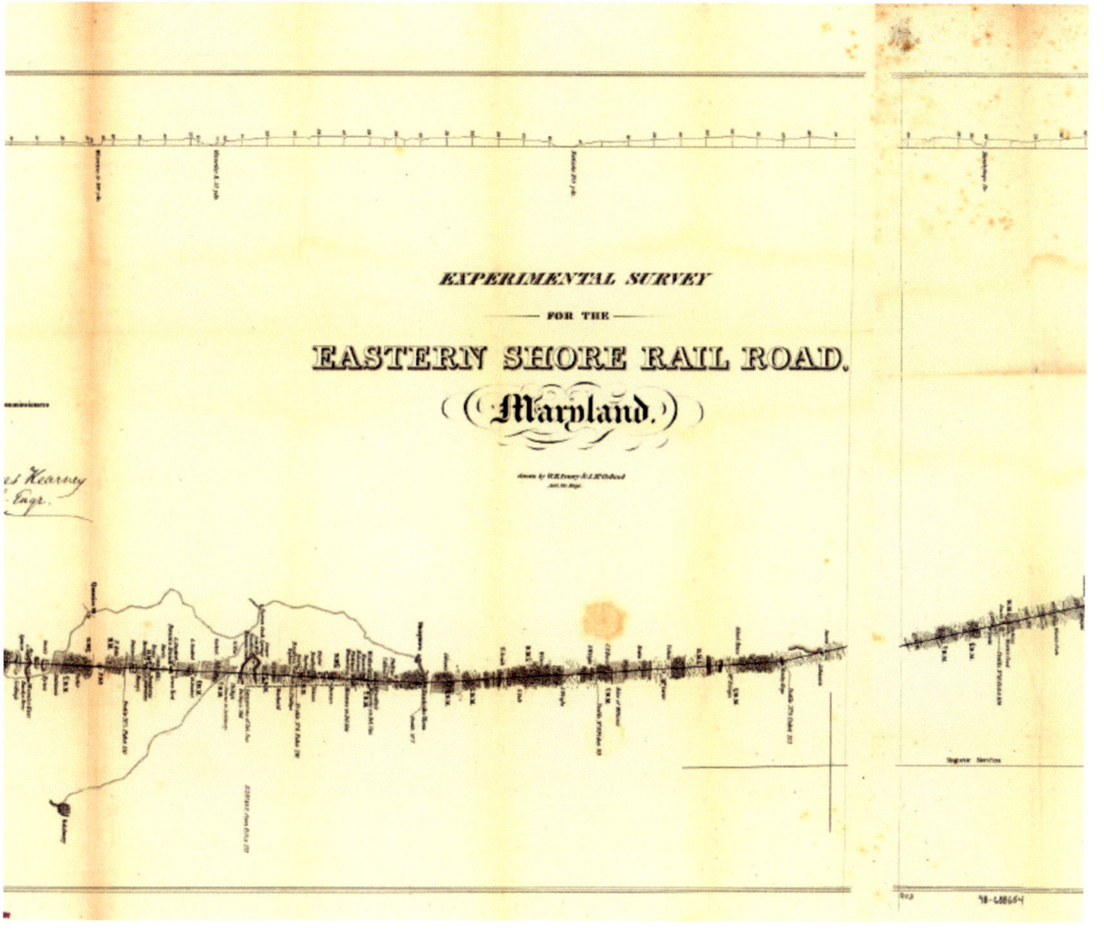

Some of the abandoned railroads on Delmarva were abandoned before the first piece of track was laid. Three of these lines tell tales that could have changed the history of the entire peninsula. The first of line not only would have changed the lines the railroad used, it would have actually changed the history of two of the states on Delmarva. In 1833, the Eastern Shore Railroad was proposed to run from a junction with the New Castle and Frenchtown near Elkton, Maryland, and travel south, with the entire rail running in the state of Maryland, ending in what is today Crisfield, Maryland. When the plan was presented, the Maryland State Legislator failed by one vote in transferring the entire Eastern Shore of Maryland to Delaware. The railroad nor the change in land from one state to the other occurred. But if one voter had been swayed, what would have become of the Mason Dixon Line?

On the previous page: Now known as the Delaware Coach Company, trackless trolleys were first introduced in Wilmington on September 24, 1939. Constructed by ACF-Brill Trolley Coaches of Philadelphia Pennsylvania, these buses increased access to the city because all that was needed were the overhead power lines and not the tracks required by the previous trolley system. Maintenance cost and downtime were decreased by using these units, so much so that on any given day, 97 percent of the fleet was in operation. Due to traffic congestion and more people buying cars and moving to the suburbs, the trackless trolleys days were numbered. The final trolley ride in Delaware was December 6, 1957. For years, the buses sat in a city junkyard with hopes of selling the units off. Today, three of these units are located at the Seashore Trolley Museum in Kennebunk Port, Maine.

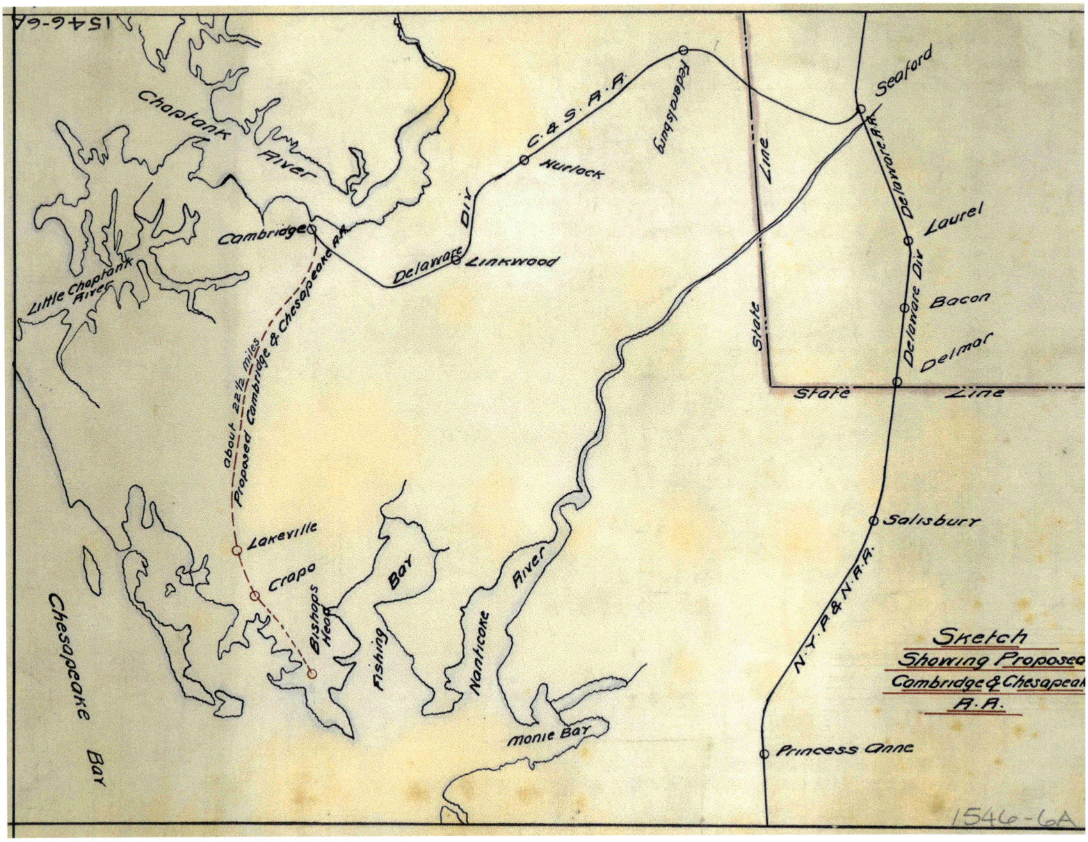

In August of 1892, a group of men proposed a rail line from Cambridge, Maryland, to Bishop Head, Maryland. The area around Bishop Head was rich in seafood of all kinds. Leaders of the railroad and local fisherman envisioned a "Seafood Metropolis," the largest in Maryland. Their county neighbors to the north had farmland and orchards that grew some of the best crops in the Mid-Atlantic area. The 22-mile line would allow these desired commodities to market within hours instead of days. The line, however, struggled to raise the needed monies to begin any construction. On three separate occasions, the last in 1912, locals tried to get the line financed. The seafood industry, as well as the entire history of the region, would be vastly different if this short line had been constructed.

On the next page:

Above: Vacation destinations were a big part of where the railroads lines were constructed on Delmarva. While freight was always the money maker, passenger trains and the revenue that would come with them were vital to the railroad's overall profitability. One such vacation spot was Betterton, Maryland. Founded the seventeenth century, the city once saw no less than eleven steamboat landings a day with visitors from Philadelphia, Baltimore, and Washington, D.C. With no real freight along a proposed line, a trolley line instead was proposed from Still Pond, Maryland, to Betterton. Sadly, the trolley line was never built and slowly the city beach area fell into decay. But that was not the end of their story. The city purchased the beach and cleaned the area. The city is now home again to tourists and a beautiful beach community.

Below: Today, places like Eastville, Virginia, sit in silence. The trains are now gone. This area is currently being evaluated for removal of the tracks, ending an over 100-year railroading history. Industries have moved, some overseas. Large produce growing and canning are now mostly done out west. Slowly, the land is reclaiming the rail bed.

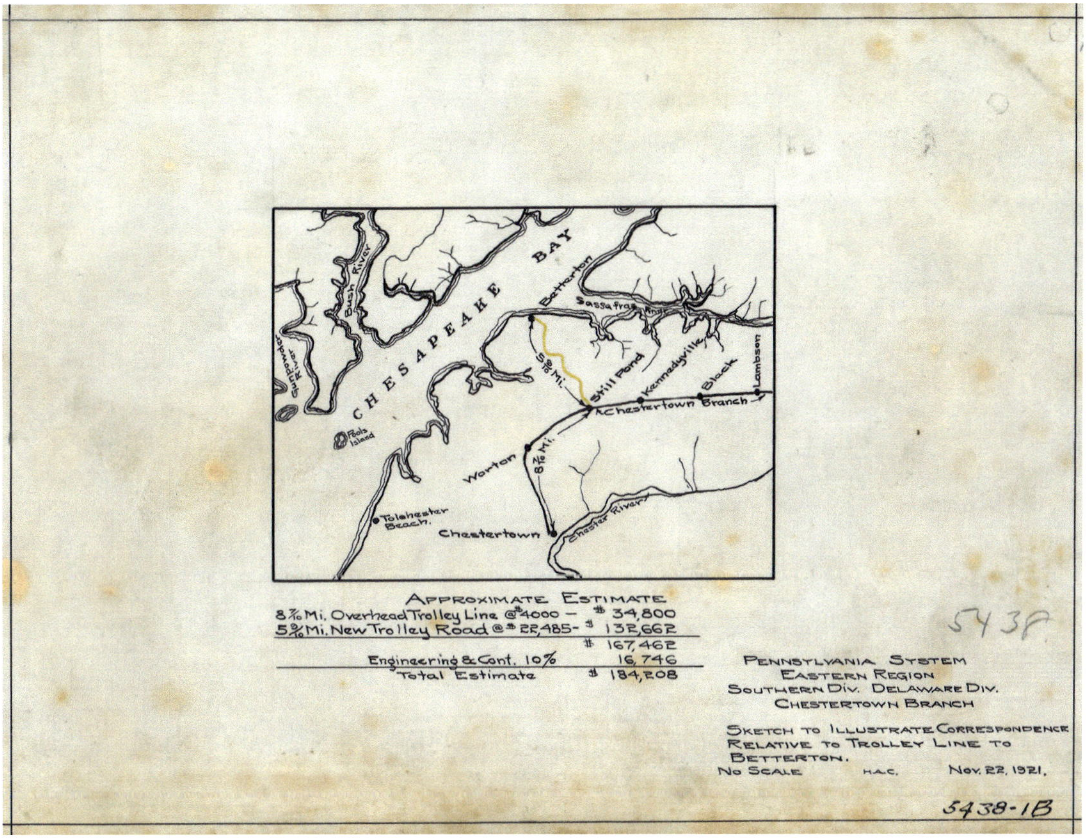
CHESAPEAKE BAY
Bush River
Gunpowder River
Pools Island
Betterton
Sassafras River
Still Pond
Kennedyville
Black
Lambson
Chestertown Branch
Worton
Tolchester Beach.
Chestertown
Chester River
APPROXIMATE ESTIMATE
8 7/10 Mi. Overhead Trolley Line @ $4000 – $ 34,800
5 9/10 Mi. New Trolley Road @ $ 22,485 – $ 132,662
$ 167,462
Engineering & Cont. 10% 16,746
Total Estimate $ 184,208
5438
PENNSYLVANIA SYSTEM
EASTERN REGION
SOUTHERN DIV. DELAWARE DIV.
CHESTERTOWN BRANCH
SKETCH TO ILLUSTRATE CORRESPONDENCE
RELATIVE TO TROLLEY LINE TO
BETTERTON.
NO SCALE H.A.C. NOV. 22, 1921.
5438-1B

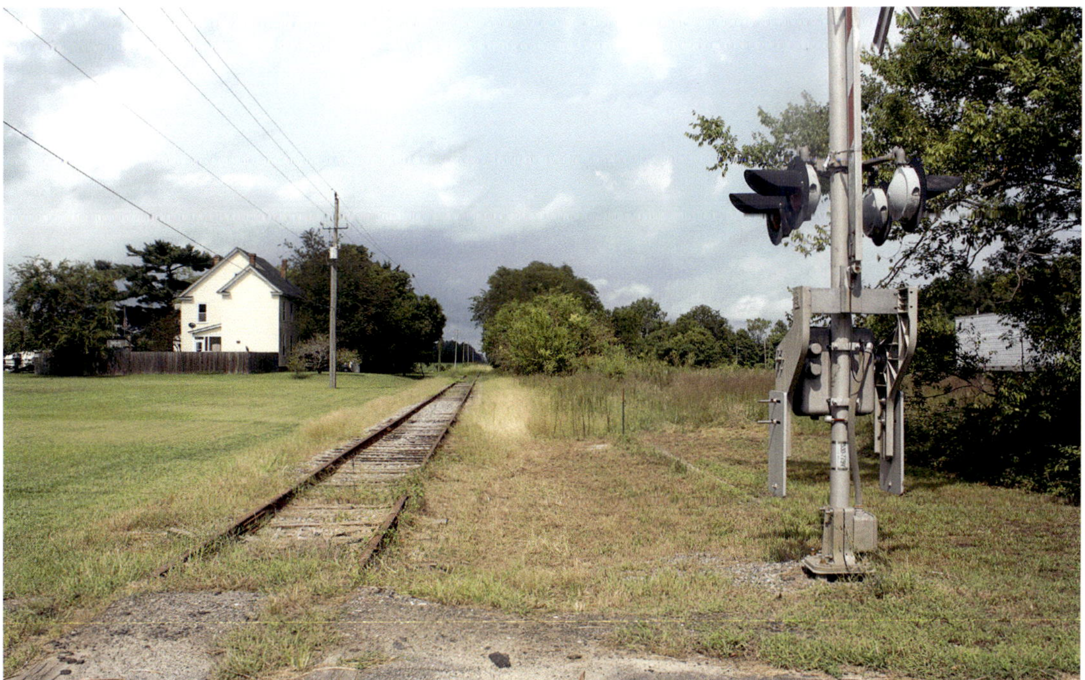

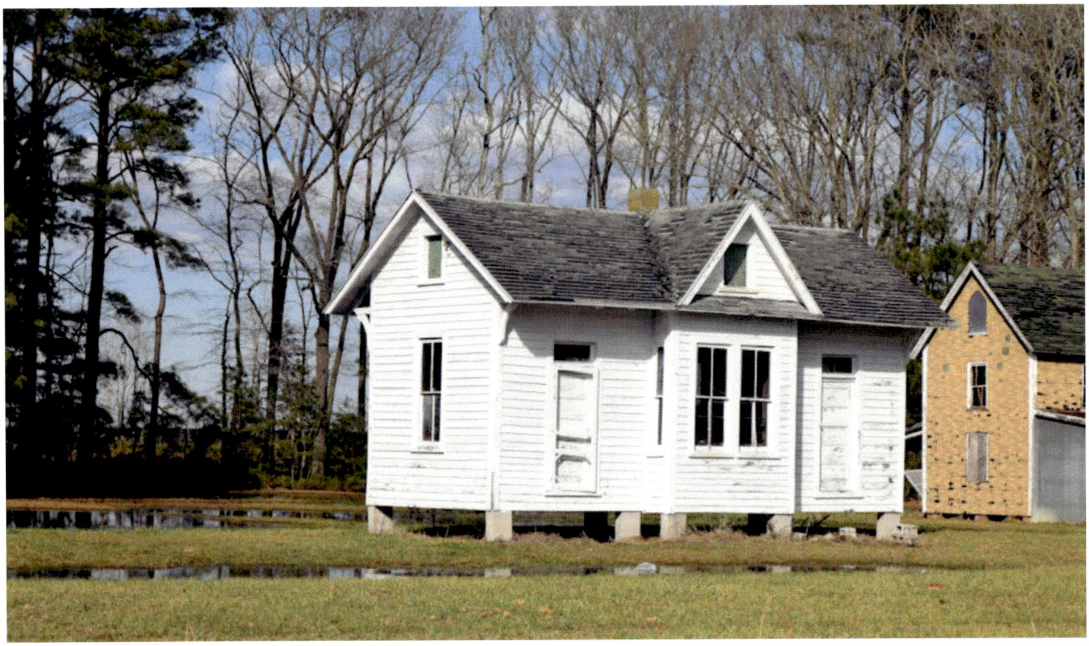

Stations such as Westover, Maryland, sit vacant and forgotten. Moved from the railroad area, many have been left to decay, people riding past paying little attention to the building's history. Most stations are gone, torn down, and the lumber reused. Railroad buffs today can visit where the stations were with occasionally a concrete curb the only remaining marker showing where these buildings once sat.

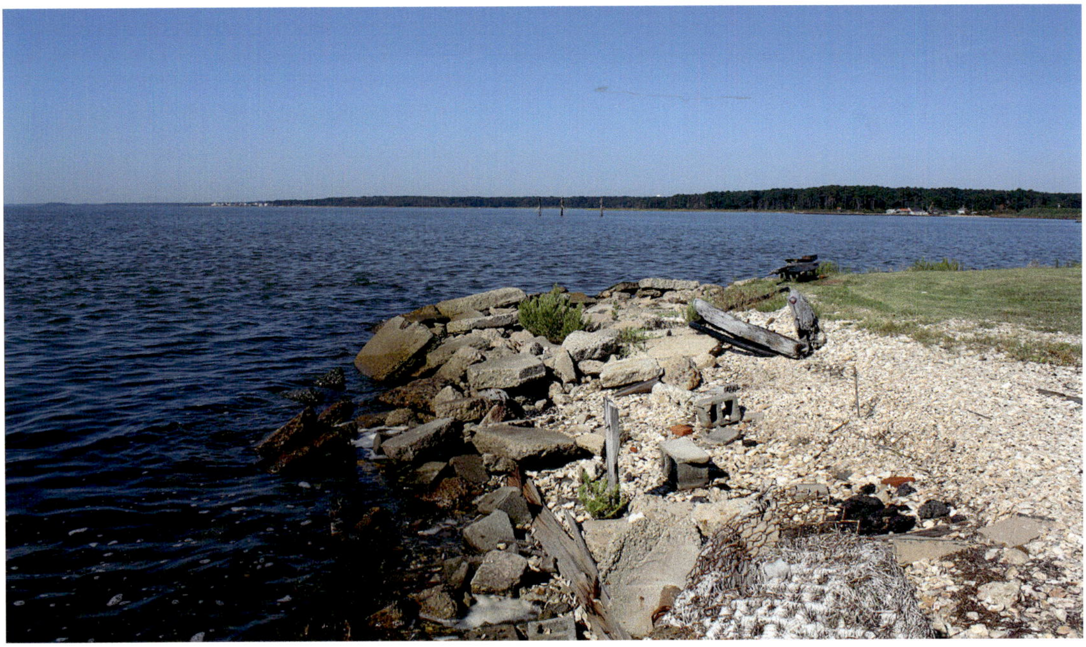

Along the water's edge in Franklin City, Virginia, shows a scene you can find at any landing that once met the railroad. As with so many, Franklin City had multiple wharfs, seafood plants, and shipping to the island of Chincoteague. Now the water has reclaimed the wood, concrete lays strewn across the once-busy port. The city here has virtually disappeared, and the old station has been approved for destruction.

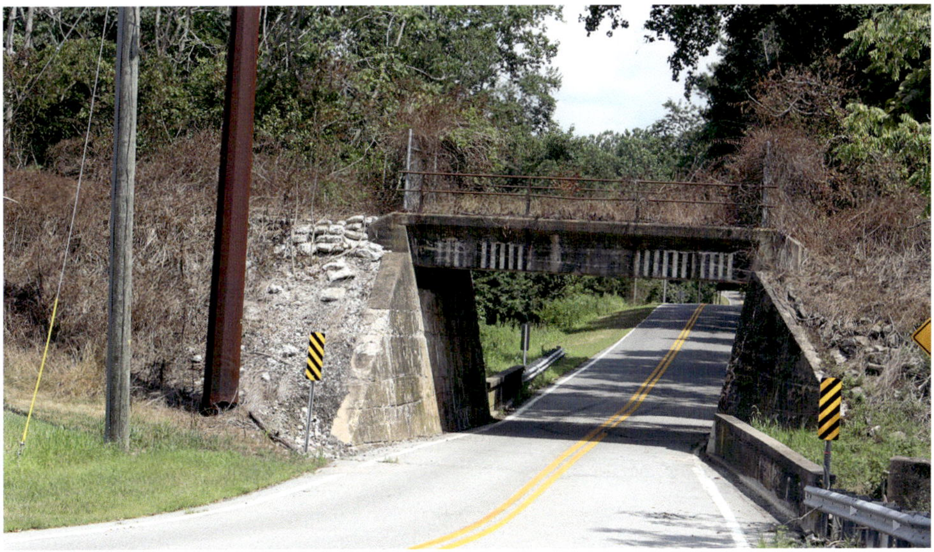

The infrastructure that once carried multiple trains a day now sits forgotten and allowed to decay. The overpass near Cordova, Maryland, has been closed with chain link fencing that has now become vine covered. The last train rolled through the area in 1993, following years of declining service in the area. While the Maryland Transit Authority reviewed the line for possible historical preservation, the line was ultimately abandoned and has been left to either be removed as needed or degrade in place.

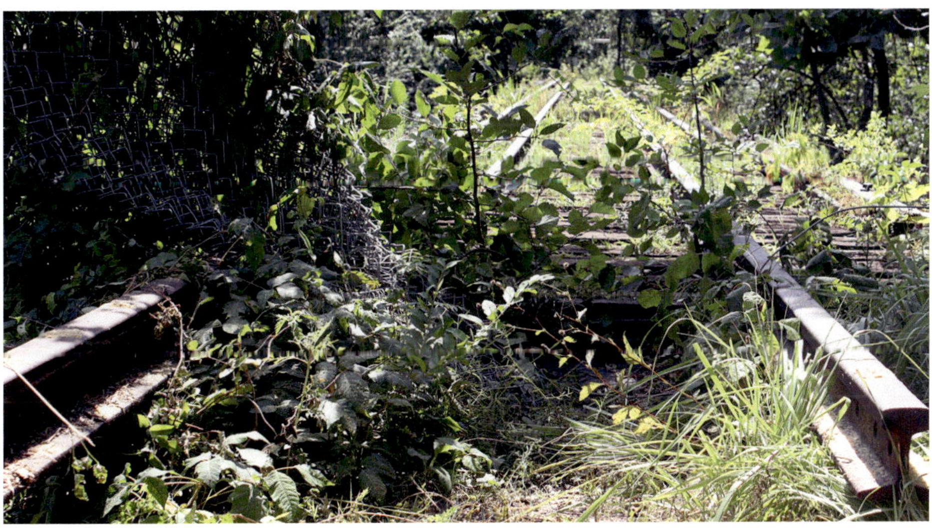

From 1865 to 1916, the "golden age" of railroads was like no other. For over half a century, no other mode of transportation challenged the railroads. During these years, the rail network grew from 35,000 miles to a peak of 254,000 miles in 1916. Economic downturns and the rise of the truck caused rail lines to slowly shrink. Today, just 92,837 statute miles remain, but that does not tell the whole story. Freight by train is still the number one method chosen in America and passenger service has seen a recent revival with such projects as those in Florida. While so many of the rails across Delmarva have been abandoned, one only needs to stop and look along the roadside, and you will probably see remnants of the past. Ride through a small town and look for the history of the railroad. If you grab a soda and sit nearby, close your eyes and you can hear the clickety-clack and imagine the hustle and bustle of the station. The rails are never really abandoned as long as they are not forgotten.

About the Author

Douglas Poore has been a longtime railroad buff. Born and raised in Harrington, Delaware, he grew up with the railroad in his backyard. Some of his earliest memories were anxiously awaiting the James E. Strates Shows' carnival train arriving each year in July, bringing the carnival rides to the Delaware State Fair. From an early age, he also owned model railroads. From his grandfather's old Lionel, to an HO scale layout that took up over half of his parent's attic, Doug has felt the passion of steam engines running in his veins. After retiring from a paramedic career, he took his love of history and railroading and became the volunteer curator of the Greater Harrington Historical Society.